FLATTERLAND

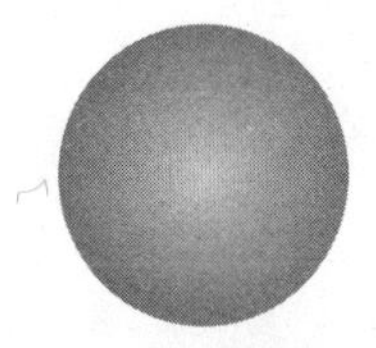

LIKE **FLATLAND**, ONLY MORE SO

A CIP record for this book is available from the Library of Congress.
ISBN 0–7382–0675–X

Perseus Publishing is a member of the Perseus Books Group.
Find us on the World Wide Web at http://www.perseuspublishing.com

Perseus Publishing books are available at special discounts for bulk purchases in the U.S. by corporations, institutions, and other organizations. For more information, please contact the Special Markets Department at the Perseus Books Group, 11 Cambridge Center, Cambridge, MA 02142, or call (617)252–5298.

First paperback printing, March 2002
4 5 6 7 8 9 10—04 03 02

CONTENTS

FROM FLATLAND TO FLATTERLAND

Sometimes writers get a bee in their bonnet – an idea that buzzes around for years until one day it suddenly crystallizes. Yes, it's easy to crystallize a bee: you just have to get the right mix of metaphors. *Flatterland* is a crystallized bee. Let me tell you how it came about, and why.

In 1884, in Victorian England, a headmaster and Shakespearean scholar named Edwin Abbott Abbott – that's right, two Abbotts – wrote a classic of scientific popularization called *Flatland.* Written under the pseudonym 'A. Square', it tells of a world of two dimensions, a flat Euclidean plane that came straight out of the geometry texts of that period. Abbott would have used them in his school. The inhabitants of Flatland are geometric figures – lines, triangles, squares, pentagons . . . The rather narrow Victorian attitudes of A. Square are shattered by rumours of the Third Dimension, confirmed by a visitor from that extra-dimensional realm who is named The Sphere.

Flatland reprinted within a month and has never been out of print since. Its appeal has survived all intervening scientific and social upheavals. It exists in numerous editions, and several writers have published sequels or derivative works, such as Dionys Burger's *Sphereland* and Kee Dewdney's *The Planiverse.*

Flatterland is another.

The scientific purpose of *Flatland* was serious and substantial. Abbott's sights were focused not on the Third Dimension – familiar enough to his readers – but on the Fourth Dimension. Could a space of more than three dimensions exist? Where would you *put* it? Abbott softened up his readers' resistance to such an outlandish notion by making them imagine how a Flatlander would respond to the outrageous suggestion that a Third Dimension could exist.

He had a second purpose, a very different one: to satirize the rigid social structure of Victorian England, with its hierarchies of status and privilege – especially the lowly status accorded to women. To this purpose he made the females of Flatland mere one-dimensional lines, inferior even to the slimmest of isosceles triangles, and vastly inferior to the circular Priesthood. *Flatland* was – and still is – a very subversive book. Some supporters of female emancipation misunderstood Abbott's satire, and in the preface to the second edition he was forced to explain that A. Square

> has himself modified his own personal views, both as regards Women and as regards the Isosceles or Lower Classes . . . But, writing as a Historian, he has identified himself (perhaps too closely) with the views generally adopted by Flatland, and (as he has been informed) even by Spaceland Historians; in whose pages (until very recent times) the destinies of Women and of the masses of mankind have seldom been deemed worthy of mention and never of careful consideration.

So there! Abbott was, in fact, a social reformer who believed in equal educational opportunity for all social classes and genders; Flatland's narrow-minded social system is his cry of frustration.

The Fourth Dimension was hot intellectual property in Abbott's day. His interest in it came a decade before H. G. Wells's celebrated science-fiction story *The Time Machine*, which was serialized in the *New Review* (1894–5) and published as a book by Heinemann in 1895. Wells, like any good science-fiction author, based his tale on a solid dose of scientific gobbledegook, which in this case is supplied by the Time Traveller:

> 'But wait a moment. Can an *instantaneous* cube exist?'
>
> 'Don't follow you,' said Filby.
>
> 'Can a cube that does not last for any time at all, have a real existence?'
>
> Filby became pensive.

> 'Clearly,' the Time Traveller proceeded, 'any real body must have extension in four directions: it must have Length, Breadth, Thickness, and – Duration . . .
>
> '. . . There are really four dimensions, three which we call the three planes of Space, and a fourth, Time. There is, however, a tendency to draw an unreal distinction between the former three dimensions and the latter, because it happens that our consciousness moves intermittently in one direction along the latter from the beginning to the end of our lives . . .
>
> '. . . But some philosophical people have been asking why *three* dimensions particularly – why not another direction at right angles to the three? – and have even tried to construct a Four-Dimensional geometry. Professor Simon Newcomb was expounding this to the New York Mathematical Society only a month or so ago.'

Wells's reference to Newcomb reminds us that in the late nineteenth century higher-dimensional geometry was all the rage among mathematicians, especially Arthur Cayley (1821–95) and J. J. Sylvester (1814–97). Sylvester emigrated to the States and became a major founding figure of American mathematics. Some of the ideas about higher dimensions were introduced into physics by Hermann Minkowski (1864–1909), and were among the things that led to Relativity, the brainchild of (among others) Albert Einstein (1879–1955).

I'm giving you the dates to make it clear that *Flatland* is slap in the middle of all this. Abbott was a prolific writer, with some sixty books to his credit, though none of the others are remotely similar to *Flatland.* He was a passionate educator, and evidently understood the difference between serious science and solemn science. His fascination with some of the cutting-edge science of his day, in conjunction with these attributes, gave the world a classic.

At the start of the twenty-first century, mathematics and science have moved a long way from where they were at the end of the nineteenth. The Fourth Dimension is mild indeed compared with the mind-boggling inventions of geometers and physicists – spaces with infinitely many dimensions, spaces with none, spaces with fractional dimension, spaces with finitely many points, curved spaces, spaces that get mixed up with time, and spaces that aren't really there at all. The *respectable* reason for allowing that bee in my bonnet to crystallize is that there is ample scope to play Abbott's game again in a new context – indeed, in lots of new contexts.

I must confess, however, that there is a less respectable reason too. I'd been toying with writing some sort of sequel, and I'd got it into my head that since *Flatland* was about the adventures of A. Square, the update ought to follow the adventures of one of his modern descendants. We know that A. Square had children, because Abbott tells us so – but he doesn't tell us what the 'A.' in 'A. Square' stands for, and that bothered me. At this point, my thoughts were distracted by the sudden realization that I *knew* what Mr Square's first name must have been.

Obviously, the 'A.' stands for 'Albert'.

British readers will immediately appreciate why this has to be so. There is a popular TV soap called EastEnders – I'm not a fan myself – I don't like soaps – but things seep out into the general 'extelligence', and EastEnders is one of them. Anyway, this particular soap is set in a fictitious region of London centred around 'Albert Square'. There is no real Albert Square in central London, but there is an Albert Hall, an Albert Memorial, and an Albert Embankment. Albert was Queen Victoria's consort, and she loved him dearly, so London is littered with Alberts. Abbott was writing in Victorian times, satirizing Victorian values . . . the name *fitted.*

Now, there may not be an Albert Square in real central London, but there is a Grosvenor Square, a Berkeley Square (in which, according to the song, a nightingale sang . . .) and a Leicester Square. Suddenly the family tree of the Square dynasty was falling into place – Grosvenor, Berkeley, Lester . . .

What of the womenfolk? On Flatland, women are lines – and it followed just as night followed day that my central character should be female, and her name should be 'Victoria Line'. Vikki for short. There are Victorias all over London, too, including a main-line train station. The Victoria line is an underground railway line (the 'tube', or subway) linking Victoria Station to Euston Station, another main-line train station. So Victoria Line it was. Another underground line, the Jubilee line, gave me Vikki's mother, Lee.

Crazy, but it crystallized the whole bee into a book. Everything came to me in a rush. Young Vikki is Albert's great-great-granddaughter, a thoroughly modern young woman in a society rather like Britain and the US in the early sixties, but with a dash of Internet thrown in for narrative lubrication. Flatland's male-dominated Victorian culture is beginning to crumble as its women break away from their traditional restraints. Vikki finds an old notebook, Albert's original manuscript of *Flatland*, and is bitten by the 3D bug – much to her parents' consternation. Unknown to them, she tries to follow in her ancestor's vertex-steps (squares have vertices, but no feet) into the expanded universe of the Third Dimension . . . and she succeeds.

Now all I needed was to equip her with a guide. Dante had his Virgil to conduct him through the *Inferno*; A. Square had his Sphere . . . but Vikki would need a more versatile guide, familiar with dozens of mathematical and physical spaces . . . For weeks I grappled with this problem, until one evening I remembered a children's toy – a rubber-skinned inflatable orange ball with horns, plus two eyes and a broad grin painted on the front. The child sits on the ball, grabs the horns, and bounces across the floor. The toy was called a Space Hopper (the grin is that of an alien from Outer Space). Evidently, a Space Hopper is ideally suited to hop from one mathematical space to another . . . and I had my guide.

After that, all I had to do was choose *which* spaces were important enough to include, and I did that with an eye on the frontiers of today's mathematics and physics. Oh, yes – I also needed a title. I hope you can excuse the train of thought that led me to choose what now seems inevitable for a *Flatland* sequel – *Flatterland.*

Please pardon my temerity in attempting this add-on to a classic. You will see at once that I can't use Abbott's elegant Victorian prose style – modern times demand modern cadences. The original, fortunately, is untainted by my irreverence (re-read your copy: I promise that not a word has changed!).

I.S.

Coventry, July 2000

THE THIRD DIMENSION

Seen from space, it was a strange world, with the austere beauty of a page from Euclid. In fact it *was* a page from Euclid, geometry made flesh: a sprawling, humming world of two-dimensional shapes. Flatland. A land of lines, triangles, squares, polygons, circles . . . people, of their own kind. They lived polygonal lives, ate polygonal food, drank polygonal drink, made polygonal love, bore polygonal children, and died (polygonally) in a two-dimensional universe – and never thought it the least bit curious. Their flat world was all they could see, all they could hear, all they could feel. To them, it was all there was.

As long as nothing disturbed that perception, it was *true.*

But times were changing in Flatland.

•

The house was dowdy and unfashionably pentagonal, but in an excellent location: just along the street from the Palace of the Prefect. It had been in the Square family for almost 150 years, and was now beginning to show its age. Nonetheless, it was a comfortable dwelling, with the typical large Flatlandish entrance hall, seven rooms for the male members of the household, two apartments for the females, a study, a large room that once had housed servants but was now used as a kitchen, with a dining alcove, and a musty, cluttered cellar. It had separate doors for women and men – for safety reasons, women being rather sharp if encountered end-on. In the hall a middle-aged woman swept up after her two untidy square sons and her neat and lineal daughter, waving her body from side to side so that the males wouldn't accidentally blunder into an endpoint and cut themselves. She found it a comfortable life, though hardly a fulfilling one, and on the whole she was content with her lot.

In the cellar, her daughter Victoria was anything but content with hers. Flatland was a sexist subtopia in which women, seen by their menfolk as simple-minded one-dimensional creatures, performed only menial tasks. Even the lowest of the males, the isosceles triangles, had higher status, and each generation of males made sure everything stayed that way. Not exactly deliberately . . . well, not consciously . . . well, not with *malice*, anyway – they really thought it was the only option . . . Well, most of them. It just never occurred to the men that Flatland society might order itself differently. And it certainly never occurred to them that their most cherished beliefs about Flatland society might be based on prejudice and unchallenged assumptions. How could it be? In Flatland, your position in society was determined by how many sides you had and how regular your perimeter was. It was an objective test, hence unquestionable.

At the top of the tree were the Circles, priesthood-cum-nobility: glorious, almost transcendent beings – perfection made flesh. And the biggest bunch of snobs you could imagine. They weren't even true circles, just polygons with an awful lot of sides. Like many aspects of Flatland, their name was a polite fiction. Behind the rigid façade of Flatland society, however, the winds of change were starting to whine. They had begun as a gentle breeze when the Six-Year War between the Axials and the Alignment had thinned the ranks of Flatland males and thrust women into the munitions factories and the civil service. To the surprise of the men, and the quiet satisfaction of the women, the lineal ladies carried out what had previously been men's jobs with aplomb – maybe too much aplomb. There were mutterings in the Halls of Power – but the catenary was out of the bag, and no amount of effort would ever get it back again. As the decades passed, the breeze had stiffened to a howling gale, as the advance of technology brought with it inevitable social spin-off.

If Vikki Line had her way, the gale would soon become a roaring hurricane. Not that she disliked boys, you understand – as long as they knew their place. Once they stopped flaunting their vertices and comparing how many edges they had, some of them were even quite nice. In fact, that was what had brought her into the dusty recesses of the cellar: she was hoping to find some old discarded clothes of her mother's to wear to the disco that evening. Roger Rectangle was taking her on a date, the retro look was all the rage, and she was hoping to find a few items that would put a kink in the other girls' endpoints.

So far, all she'd found was a motheaten dishcloth and a box of her father's old string vests. (Literally: most Flatland garments were flexible lengths of string, which Flatlanders wrapped round themselves and secured with sticky tabs, leaving their faces uncovered.) Vikki had a feeling that her mother Jubilee would throw a fit if her daughter tried to go out wearing a string vest, however chic.

She noticed a cluster of rectangular boxes in a corner. One of them was battered and fraying at the corners, which looked promising; but she couldn't reach it easily, so she tugged at one of the nearer boxes. With a crash, the mildewed cardboard disintegrated and the contents scattered across the cellar floor.

'Victoria? *Victoria!* What was that? Are you all right, dear?'

She sighed. 'Yes, Mother. An old box came to bits, that's all.'

'Oh. Well, it sounded like a herd of ellipses. Do be careful, dear. And clear up any mess you've made.'

'Yes, Mother.' Vikki started to pick up the junk that had tumbled across the cellar floor, stuffing it back into the now rather battered boxes. She had almost finished when she noticed a tattered book. (More properly, it should be described as a scroll, for on Flatland books are written on lines, not flat sheets, in a kind of Morse code; and the way to store a line compactly is to roll it into a spiral . . . I can't keep explaining this kind of thing to you, my Planiturthian readers. So if I use a Planiturthian term that seems not to make sense, for instance, having Vikki – who is a *line*, for heaven's sake – pick something up or carry something, you'll just have to assume that there is some Flatland equivalent.)

Anyway, the book, for so we shall call it, had skated behind some cracked crockery, and nearly escaped her attention.

Curiosity impelled her to roll the book open. Even in the cellar, she had no difficulty seeing it, for in Flatland light manages to make its way into every nook and cranny. Where it came from was a complete mystery – even to the greatest savants, even at the end of the twenty-first century – but its ubiquity had an architectural consequence: houses did not need windows.

The book was handwritten, in an old-fashioned script. The title page bore the words *Flatland, a Romance of Many Dimensions*. The author was identified as *A. Square*. At first she thought it was some kind of child's primer on the geography, history, and sociology of Flatland, but as she skipped through the text it began to mutate into something darker and more personal. It was almost like a diary, except that it was not arranged by date. A bit past the middle she did come upon a date, however:

> *From dreams I proceed to facts.*
>
> *It was the last day of the 1999th year of our era. The pattering of the rain had long ago announced nightfall, and I was sitting in the company of my wife, musing on the events of the past and the prospects of the coming year, the coming century, the coming Millennium.*

Why, that would make the book almost exactly a hundred years old! Vikki read on, hoping for more clues. The weird narrative told of a Stranger, a Circle who could change size – a stranger from Space. It was some kind of science fiction novel, then. A lot of the boys seemed to be into that kind of thing. A phrase caught her eye:

> *You see, you do not even know what Space is. You think it is of Two Dimensions only; but I have come to announce to you a Third – height, breadth, and length.*

Ah. They'd done this in physics. Space was two-dimensional, of course – how could it be otherwise? But there was a sense in which you could think of time as a third dimension, thereby getting a three-dimensional 'spacetime continuum'. It wasn't *real*, of course, just a mathematical invention – and she found the idea rather pointless because you couldn't draw pictures of three dimensions, anyway. Face it, you couldn't even draw pictures of *two* dimensions – you had to project real space down onto a line if you were going to draw a picture. That was how the visual sense worked. Sculpture, though – that was genuinely two-dimensional, like the tactile and auditory senses. So a *moving* sculpture could be considered three-dimensional – and that was what the Stranger in the book was. A moving sculpture that spoke: what a strange idea!

She read on, and became confused. Whatever the Stranger's three-dimensional Space was – and only now did the significance of the capital letter become apparent – it was not the conventional three-dimensional spacetime continuum from her physics lessons. In the Stranger's mind, the Third Dimension wasn't time! What, for instance, could be made of this passage?

> *You are living on a Plane. What you style Flatland is the vast level surface of what I may call a fluid, on, or in, the top of which you and your countrymen move about, without rising above it or falling below it.*
>
> *I am not a plane Figure, but a Solid. You call me a Circle; but in reality I am not a Circle, but an infinite number of Circles, of size varying from a Point to a Circle of thirteen inches in diameter, one placed on the top of the other . . .*

Above? Below? On *top*? If A. Square had been writing about the spacetime continuum, surely he would have used words like *before, after*. These terms – she could try to infer their content from their context, but the inference didn't really work – were outlandish, meaningless nonsense words.

Vikki Line tucked the curious book away inside her edgebag, an accessory that had become very fashionable indeed, and not only among the young. Even males were wearing them, though they called them 'sidesacks' to distinguish them from effeminate edgebags. She would take a closer look at the book later on. Right now there was a more pressing problem: finding something suitable to wear to the disco.

•

'I don't understand why you young people insist on wearing your parents' old cast-offs', her mother fussed. 'You know, your father gave me that jacket just after we got married. It used to fit me then. The colour suits you, dear, I must say.' Only a few generations ago, Jubilee would not have been able to make such a statement – the rule had been 'any colour as long as it's grey'. But the old colour prejudice, fallout from the political sabotage of the Universal Colour Bill, was slowly dying out (indeed, dyeing out), except for a continuing prohibition on body-paint – and even that was coming under fire with the new fashion for tipstick among upper-class young women.

'It really needs an iron, though, dear,' Jubilee fussed. 'Would you like me to—'

'New clothes are so *ordinary*, Mother. I want to look different.'

'Different from what, dear?'

'Well, just . . . different. Like all my friends.'

'Different but the same, then, dear?'

'Ohhhh! You're making fun of me again!'

Her mother smiled (Flatland women do this by wiggling their front vertex in a special way). Vikki took the treasure away to her own little room, along with her other discovery. She kept her clothes in there, and her personal belongings – a Parallelogram Bear and a dilapidated My Little Polygon which she had long ago outgrown but kept for sentimental reasons; a tape-player with hundreds of cassettes; letters from her friends; schoolbooks; and – her pride and joy – one of the new Home Computers, complete with key-strip, tape-reader, printer, and scanner. It had a 2D graphics accelerator, and twenty megs of RAM. Bundled with it had been a 'free gift': a tiny electronic Personal Disorganizer, in which she kept contact information for her friends and an extensive diary. It communicated with the computer by invisible light. Not only that – once she'd saved enough money she was going to add a modem, persuade Daddy to rent an extra phone link, and surf the InterLine. It wouldn't be hard to persuade him: all she had to do was to stay on the phone to her friends for hours and burst into tears and tell him he was ruining her life if he dared interrupt to make his own calls. Her friend Dilly had tried it on *her* dad a month ago, and a second phone connection had been installed within a day.

The Flatland phone system, by the way, was a triumph of technical ingenuity. In a two-dimensional world, you can't lay a network of cables without trapping people between them – there's no underground and no overhead. But you can avoid cables altogether and use radio waves of a frequency to which most things in Flatland, especially houses and people, are transparent. With enough repeaters scattered around to boost fading signals back up to full strength and shunt messages past radio-opaque objects, the system worked surprisingly well. Fortunately, Flatlanders seemed to be unaffected by the radiation that sleeted through them, though some consumer groups were beginning to worry that overexposure to the phone system might be contributing to an epidemic of inflamed centroids.

At the very moment that Vikki was thinking about her father, he arrived home from work. Grosvenor was a huge, good-natured square . . . well, actually he had gone a bit trapezoidal in his middle age, mostly because Jubilee was such a good cook. As usual, he had picked up their young sons Berkeley and Lester from primary school on the way home. Grosvenor gave Jubilee and Victoria a homecoming kiss and flopped against the big sofa in front of the fire. The soft, slightly springy cushions closed snugly around three of his edges. (The distinction between sitting on/in/against a piece of furniture and *wearing* it was often rather fuzzy in Flatland.)

The boys shot off into the yard, to play until dinner was ready.

'Vikki, love, be a dear and bring me a beer, will you? It's been a bisector of a day at work.'

Typical. But thoughts of an extra phone link stopped her saying what was on her mind: *get your own beer, Dad.* Instead, she padded off obediently and brought him back a rectangle of lager from the freezer.

He popped the tab at one end and sucked. 'Thanks, sweet-centre.' *Well, at least he's expressed gratitude.* 'Like the shoes. Nice jacket, too, love: it suits you.'

'It's the one you . . . I mean, yes, it does, doesn't it? I found it in the cellar.'

'I keep meaning to clean that cellar out', said Grosvenor. 'Nothing but a heap of useless junk. Some of it's been in there for generations. Give the lot to Boxfam, that's what we ought to do.'

Talk of junk reminded Vikki of her discovery. Innocently, she said, 'That reminds me, Dad: I found a funny old book in the cellar. It looks really really interesting.'

'A book, eh? Well, that cellar holds an awful lot . . . bound to be a few books down there—'

'It was handwritten, like a diary. By someone called A. Square.'

Her father sat bolt upright and popped out of the enveloping sofa. His beer slipped from his grasp and the fire hissed as a few errant drops hit it. '*What?*'

'It was all about some weird Stranger from the Third Dim—'

Grosvenor's face turned an angry shade of grey. 'Victoria Line: don't you ever mention that phrase in this house! God, I thought the family'd got rid of that pernicious little diatribe fifty years ago!'

Vikki didn't understand what she'd done wrong. 'But Dad, it's just an old—'

Jubilee, ever the calm one, touched her daughter's side affectionately. 'Wait till the boys have gone to bed, Victoria. Then your father will tell you a piece of family history.'

Grosvenor stared at his wife in horror. 'Lee, are you sure that now is the right—'

'She's old enough to know the Facts of Life, Grosvenor dear, *and* how to deal with them on a practical basis. So she's old enough to know the truth about her great-great-grandfather.'

The Facts of Life bit was news to Grosvenor, and it threw him completely. 'Dammit, Lee: great-great-grandad Albert's already caused this family too much grief!'

'Is that what the "A" stands for, Dad? Is the book by *Albert* Square? Was he my great-great-grandfather? What *is* the Third Dimension, Dad?'

'Victoria, I've just told you not to—'

'Grosvenor, it's too late. We can't hide our past from our own daughter', said Jubilee. 'And it was all so long ago. Times are changing. She has a right to know. And you *did* promise—'

Grosvenor Square sagged back against the sofa. 'Yes, but I thought she'd be a bit older than this before I had to . . . Yes, yes, I'll tell her. I'm just finding it hard getting used to having a young woman in the house instead of a little girl, OK?'

'After dinner,' insisted Jubilee, driving home her advantage, 'as soon as the boys have gone to bed.'

Grosvenor was a beaten man. 'Yes, Lee – after dinner. As you say.'

Jubilee was already dishing out the food into semicircular bowls. Vikki rushed to the door and flung it wide open. '*Berkley! Les! Grub's up!*'

•

Faint sounds of childish prattle were wafting from the boys' bedroom. Ignoring them, Grosvenor took a deep breath and tried to find the courage to peel the wraps off an ancient – and, he had hoped, forgotten – piece of Square family history.

Jubilee saw her husband was having trouble, and offered a simple solution. 'We don't talk about great-great-grandfather Albert because he died in prison, Vikki.'

'*Lee!*'

'There's no point in beating about the bush, Grosvenor. Albert *did* die in prison.'

'Yes, but he wasn't a criminal.'

'Did I say he was? Tell Vikki what he did to get himself imprisoned.'

'Uh – well, you see, Ancestor Albert . . . Lee, do I *have* to do this?'

'Yes.'

Grosvenor grunted, accepting the reality of his position. 'Very well. Vikki, Ancestor Albert was . . . he was the black shape of the family, so to speak. He . . . he got some ridiculous nonsense into his head about what he called the Land of Three Dimensions. It was an imaginary world, different from ours, and he would have been all right if it had *stayed* imaginary, but his . . . his mind went. He became convinced it was real. He claimed that he had received a visitor from the Land of Three Dimensions, which he called The Sphere.'

'What a funny word. Was the Sphere the Stranger I read about in Albert's book?'

'It was. Albert even claimed he had visited the Land of Three Dimensions himself, with the Sphere as his guide.'

'Wow! Hey, that's really neat!'

Grosvenor sighed. *The naive enthusiasm of youth . . .* 'A hundred years ago, Vikki, saying things like that got you sent to prison. For heresy, because you were contradicting the Priests, and because anyone who claimed to have visited another world must be a madman.'

'Oh.'

'So you see, sweetcentre, it's not something the family is proud of. To make matters worse, Albert's unfortunate brother was imprisoned too, supposedly because he witnessed something – a

Visitation, some such nonsense.' He paused, gulped for air, found none that helped. His voice came out half-strangled: 'Do you *really* want all the neighbours to know that two of your ancestors were lunatics?'

Vikki wasn't sure. Being imprisoned for your beliefs was kind of romantic, like being a freedom fighter. And as for witnessing a Supernatural Visitation, that was *cool*. 'Crumbs, Dad, that was a hundred years ago.'

'The taint still lingers, Vikki. If your friends found out about old Albert, you might find that some of them weren't your friends any more. I admit that people aren't as obsessed by religion as they were in 1999, but they're still unhappy about any hint of mental instability.'

You mean they're still just as narrow-minded and unimaginative as they ever were, thought Vikki. It was a sobering thought.

'Did you read Albert's book, Vikki?'

'Only a few bits and pieces, Dad. I just . . . glanced at it.'

Grosvenor sighed with relief. 'Good, at least it hasn't had a chance to taint you too.'

'It looked kind of . . . interesting. I was going to read the rest of it later.'

'*No!*' The cry was instant and automatic. 'Sorry, love, but I don't think it's suitable material for you to read – or anyone else.'

Vikki felt this was an infringement of her Polygonal Rights. 'Why not?'

'Look what it did for Albert,' said Grosvenor, a wry smile flitting across his florid features. 'Think what it might do for you, for us . . . In your room, is it? Go and get it for me, there's a good girl.'

Vikki didn't like the sound of that. 'But Dad, it's a historical document!'

'A hysterical document, more like. I'll get it, then. Where did you put it?'

Vikki gave her mother a pleading look. 'Mum! He's going to destroy it! Can't you stop him?'

Jubilee gave a negative shake of her endpoints. 'I'm sorry, but your father's right, dear. Best not to wash our dirty strings in public. What's done is done, let's not dwell on the past. You had to be told about Albert because at some stage his name might come up, and you need to know how to react. But you *don't* need to read the rubbish that put him in prison. It's not fit for a young lady, anyway.'

Oh, Mum, if only you knew some of the books I've read . . . But Vikki could tell when she was beaten. 'Wait here, Dad, Mum: I'll go and get it. Give me a few minutes, though, OK? Just to be alone with my thoughts. I promise I won't try to read any of it before I bring it to you. Trust me? Please? Give me *some* dignity?'

Her father nodded, her mother gave a smug smile. Vikki slunk out of the room, defeated.

'That was very hard,' said Grosvenor, 'I feel awful. Do you think we should—'

'She said to trust her, dear. So we shall.'

'Of course. Lee, you're so sensible about these things.'

They waited. After a quarter of an hour, Vikki was back. With a sulk, she put the book on the table in front of her parents.

Grosvenor rolled it partly open, checked the title, sampled a few lines here and there to be sure it was the authentic copy. There was trust – and there was *trust*. 'Should have been burned long ago,' he said. Then he tossed it into the fire. 'Time you got yourself ready for young Roger, the lucky dog. Go out and have some fun.'

Vikki watched, damp-eyed, as the flames turned her great-great-grandfather's life's work into smoke and ashes. 'Anything you say, Father.'

2

VICTORIA'S DIARY

Wunday 25 Septober 2099

It's been a day of lefts and rights, Dear Diary, and I'm still reeling from the shock. I have a – get this – infamous ancestor! Isn't that absolutely ORTHOG? My great-great-grandfather Albert was a religious heretic who got himself slung in the pokey for his beliefs – and let me tell you, they are *wild*!!!!! I hardly know where to BEGIN!!!!! Basically, they're about . . .

Victoria stopped, unsure of the best way to convey the essence of Albert's heresies to the encrypted Diary in her Personal Disorganizer.

After Grosvenor had burned Albert's book – *what about freedom of speech, Dad? What about freedom from censorship?* – she had fled to her tiny room, snuggled up against the thin cushion that served as the Flatland equivalent of a bed, and sobbed her heart out until she heard her father tipvertex away from the door of her room. She had continued sobbing for a few minutes longer, letting the level fade, until finally she stopped. Then she had put on a bit too much tipstick, found a skirt that was just a little more daring than Dad normally allowed her to wear to discos, slipped into Mum's wonderful – though, she now noticed, slightly smelly – old jacket, grinned wickedly, snivelled until her mascara ran, and emerged into the communal hall for an inspection from her mother. Jubilee noticed the streaks of mascara, admonished Grosvenor for upsetting his sensitive little daughter, and hauled young Victoria back into her room to do a proper repair job on her lineliner. Ten minutes later Grosvenor was still looking sheepish, and Roger had arrived to escort her across town to the disco as arranged. The skirt was not mentioned: Grosvenor knew when he was on a hiding to nothing.

It had been a successful evening, and Roger brought her home again with five minutes to spare. Best to show willing. She said goodnights to her parents, and retired to her room.

Within seconds she had switched on the Personal Disorganizer and accessed her Diary. There was so much to tell before she forgot.

> . . . Basically, they're about a new kind of universe. That's right, a whole new UNIVERSE. And what thrills me to the core, Diary Dear, is that Old Al actually WENT there!!!! Of course, Dad made me hand Al's book over to the Authorities – namely, HIM – I should've known he'd do that, the ancient book looked much too interesting for me to be allowed to look at it, let alone KEEP it. And, entirely predictably, he burned it.
>
> Fortunately, Dad doesn't know how to work my computer, and he certainly doesn't know how to crack Most Excellently Private Encryption, so I am free to reveal to you a few things of which my dear Papa knows not one whit. (MEMO: find out what a whit is.) I did promise not to read Al's subversive pamphlet during the fifteen minutes that Mum and Dad left me alone in my room to console myself about the impending loss of major historical data, and you will not be surprised to hear that I KEPT that promise, me being a good little butter-wouldn't-melt-in-her-trisectors young lady of horrendously decent upbringing.
>
> However, I didn't promise not to . . .

Victoria broke off typing. Best to check if the file was intact before looking foolish in front of her own Diary.

She booted up the computer and checked.

It was.

> . . . I didn't promise not to read it LATER. Now, you may well ask: how could I read a book that Father had burned? (Go on, humour me.) Ah, since you ask, Diary Dear: that is the oh-so-very clever part. You see, instead of spending those vital fifteen minutes snivelling about having to give way to Higher Authority, as Father

no doubt imagined I was doing, I spent them scanning Al's magnum opus into my cute little computer. I now have two copies in the machine, three more backups on tape, and yet another backup in my Personal Disorganizer, so Albert Square's seditious screed is safe for posterity!!!! (meaning me). I now intend to read it from vertex to vertex, and I shall faithfully record my discoveries in YOU, Diary Dearest. But be patient with me, for I must make sure my parents don't catch me reading it, otherwise there will be convex hull to pay. OK?

Now, to more personal things. Let me tell you about racy Roger!!!! He really is one of the cutest quadrilaterals I've dated in absolutely AGES and he really knows how to show a girl a good time . . .

•

The days passed, and the topic of Ancestor Albert and his cranky beliefs was dropped by all concerned. Roger Rectangle fell out of favour and was replaced by Trevor Trapezium. Grosvenor felt this was a backward step, what with Trevor being even less regular than Roger, but he consoled himself that by saying nothing he was demonstrating how modern and unprejudiced he was. It was what you *were*, not your perimetral geometry, that mattered. But deep down he hoped that eventually his daughter would find a nice pentagon, get married, and present her parents with oodles of grandchildren – Hexagons, preferably – like any respectable young lady should.

The young lady's thoughts, though, were elsewhere. The lure of the illicit was proving irresistible, sucking her further and further into her ancestor's wild fantasies . . .

•

Twoday 26 Septober 2099

Dear Diary,

I've become irrationally convinced that old Albert didn't make up his story at all. I'm certain as Squares fit in the Woods that the Sphere was REAL!! My evidence is that the whole tale is *much* too imaginative to have been invented – certainly not in the intellectual climate of the previous century. Albert generally portrays himself as a boring, rather ho-hum sort of guy – which he *is*. He has enormous difficulty taking on board even the simplest 3D concepts, such as a cube. But a cube, Al my friend, is just a line's worth of squares, just as a square is no more than a line's worth of lines. That much seems evident even to me (though I suppose I'd better concede that we did the cube in physics: it's the space-time diagram of a square that persists for an interval of time). So Albert must have got his outlandish ideas from *somewhere*, and the best place to get outlandish ideas from is Outland – somewhere *external* to Flatland. And if there is such a place, why shouldn't it contain a line's worth of circles – admittedly of various sizes – which is what his friend the Sphere seems to have been?

Spheres, then, exist – if only in some Flatonic realm of higher ideals. So does *The* Sphere, I am sure – and in a far more concrete realm. 'Space', it was called in the book. And if The Sphere visited Albert, why shouldn't one of its descendants visit *me*?

The only problem seems to be attracting its attention. And that is a real pig.

Nevertheless, I'm writing computer programs to analyse Albert's treatise, in search of hidden messages. Surely he would have left some clues for others to follow in his vertexsteps.

•

As Victoria's obsession with Albert's book grew, she began to spend more and more of her time in her room, working away on the computer. Grosvenor noted this tendency with approval. 'I do believe Vikki is growing up', he confided to his wife one evening after their daughter had rushed off to her room, saying she had some urgent homework to do for a test. 'I *knew* that computer would have educational value.'

Jubilee wasn't so sure. In her experience, when a girl of Vikki's age suddenly developed new habits, that was the time to start worrying. But there was no point in disturbing Grosvenor's equilibrium unnecessarily. She would keep a quiet eye on her daughter, just to set her mind at rest.

By now Victoria could pretty much recite *Flatland* by heart. She had also learned to mistrust it, having noticed some absurdities. The most glaring was Albert's solemn declaration that from generation to generation there was an almost certain progression towards ever greater regularity. '*It is a Law of Nature with us that a male child shall have one more side than his father, so that each generation shall rise (as a rule) on step in the scale of development and nobility. Thus the son of a Square is a Pentagon; the son of a Pentagon a Hexagon; and so on.*' What utter rubbish! Why, for a start, what of her brothers Lester and Berkeley, square sons of square Grosvenor? And a moment's thought about the evolutionary history of Flatlanders, which had been going for at least a quarter of a million generations, was enough to tell you that something was wildly amiss in Albert's confident statement. For if he were correct, by now every male in Flatland would be a near-perfect Circle with a quarter of a million sides! Even the most lofty of the Priests probably had less than a hundred.

What *had* he been thinking of? She could think of a few possibilities. First, of course, the modern concept of Deep Time and a vast evolutionary history hadn't really gained common currency until a generation after Albert wrote his book, so the error wouldn't have been so glaring then. Probably Albert had taken the *aspirations* of his class and cast them as a Law of Nature – for it was as certain as anything could be that in his time ordinary Flatlanders were desperate to rise at least one level in their rigid, class-ridden society. So the occasional rare 'success' – judged by the standards of the period – had somehow become elevated to the undeserved status of Natural Law. That 'as a rule' was the giveaway. The truth – she had done this in biology some years back – was more prosaic. Most male children had the same number of sides as their fathers. A tiny proportion had fewer, usually just *one* fewer. A roughly equal proportion had an extra side; a very tiny proportion indeed had two or more extra sides. This 'random walk' of polygonal sides had, over time, led to a small number of 'Circles', while the overwhelming bulk of the population were equilateral triangles, squares, pentagons, or hexagons. This, at least, was what happened among the regular classes. With the irregulars, mainly triangles, a similar story was played out but with changes to the lengths of the sides.

Perhaps Albert had just exaggerated the truth to flatter his readers? Or was he mouthing a conventional wisdom that pretended people's hopes were reality? Certainly he was suppressing a lot of things that his society presumably found to be unmentionable in polite company: there was, for instance, no mention of starchildren – polygons so hideously deformed that their sides *overlapped*, so that a pentagonal father might have a star-shaped pentacle for a son. Colloquially known as '2½-gons', such children seldom survived more than a few years.

All this paled beside an even more pertinent observation: what about the female contribution to their offspring's genetics? Albert's male-default blinkers had prevented him from even asking the question. But even though a woman's form might be a line segment, there were clearly lines *and* lines.

And it was while lying in front of her bedroom fireplace, thumbing through the book in search of clues to this strange puzzle, that Victoria made a major discovery.

•

Threeday 27 Septober 2099

I HAVE FOUND SOMETHING!!!!!

Last night, rereading Albert's treatise for the hundred-and-umpth time, I noticed some *very* faint chicken-tracks, after the end of the book itself. The cunning old swine had written them in tiny characters and very faint ink, but my scanner is rather good at enhancing contrast. After southloading some image-processing software from the InterLine I enhanced them further, and – lo and behold – they're a list of numbers separated by dots and slashes:

18.3.1.12 / 6.12.2.90 / 2.6.2.69 / 16.1.3.20 / 20.14.1.29 / 4.1.2.15 / 19.1.3.1 / 19.1.3.2 / 8.1.1.3 / 11.2.1.28 / 12.2.1.2 / 6.13.2.40 / 18.2.1.3 / 7.3.4.28 / 12.11.1.13 / 11.2.1.14 / 14.2.1.28 / 14.2.1.18 / 18.15.2.18/ 3.1.1.1 / 18.1.4.10 / 4.1.4.16 / 4.1.3.29 / 18.20.1.6 / 6.1.2.32 / 22.2.3.15 / 22.2.3.17 / 13.6.3.21 / 13.6.3.35 / 13.6.3.16 / 21.4.2.17 / 13.6.4.15 / 1.3.2.20 / 21.7.3.58 / 21.7.3.59 / 9.4.4.16 / 16.1.3.27 / 3.3.1.11 / 3.3.3.15 / 1.2.1.11 / 5.5.2.26 / 5.12.4.13 / 7.4.3.24 / 3.3.2.4 / 3.3.1.11 / 12.3.3.34 / 8.1.1.7 / 7.1.1.20 / 14.1.1.18 / 3.3.1.24 / 1.2.1.15

and there's a good bit more in similar vein which I've stored in the computer.

It must be a coded message (cliché, ho hum). No, *not* ho hum, this is *exciting*! If Albert took so much trouble to include this funny message, it *has* to be important. And for some reason he wanted to keep it secret from the casual reader. Why? Well, I have an idea about that. After all, *I* am an anything-but-casual reader. The message is for people like me: people who are looking for truths even deeper than those recorded in *Flatland*. People so obsessed by the Third Dimension that they might just be able to find an almost unreadable message.

You're saying I just got lucky? Diary Dear, that is an extremely cynical and accurate remark, and because of it I shall stop writing this very instant. Shame on you.

•

Victoria took to spending even more time in her room. Jubilee, growing ever more suspicious but having no idea what she should be suspicious *about*, searched the room every day while her daughter was at school, but found nothing untoward. Though there were lots of scribblings with numbers and letters all over them – mathematics homework, by the look of them.

•

From Vikki's Notescroll directory

- It can't be an alphabetical code, because it includes numbers like 90, and our alphabet only has 73 symbols.
- I've tried all of the decryption software that I can southload from the InterLine, without result. It's not a transposition cipher, nor is it a polyalphabetic one. And public-key crypto wasn't invented in Albert's day, so it's not that, thank God. It can't possibly be a one-time scroll: it *has* to be decipherable, or else Albert wouldn't have put it in his book. But it's only intended for people who are willing to put in plenty of effort to decode it, that's for sure.
- The numbers break into groups of four: 18.3.1.12, then 6.12.2.90, and so on. Why?
- The groups involve different ranges of numbers. The first in each group only goes up to 22. The second is usually small (like 2 or 3) but occasionally goes as high as 15. The third is

always small: the highest is 4. The fourth, though, sometimes gets up towards a hundred. This *has* to be significant.

- It must be based on some *key*. Probably a standard text. I have tried *Polygon With the Wind*, *Parallels Lost*, *A Tale of Two Circles*, the *Complete Works of Shakesquare*, and sundry religious texts – all without success. What book would Albert have chosen? I am baffled.
- Idiot. There is one obvious book – *Flatland* itself. But what do the numbers *mean*?
- *Flatland* has 22 chapters. The first number in each group never exceeds 22! I do believe . . .
- *Yes!* It has all become clear. The first number in each group of four refers to a chapter, the second to the numerical position of a paragraph within that chapter, the third to a sentence within that paragraph, and the fourth to a word within that sentence.

How crass! Anyway, after a few recounts I have cracked the code, and the SECRET MESSAGE reads:

> O Learned traveller who aspire to follow me the method is perplexing and perilous do not reveal your intentioned The Sphere can be summoned by drawing diagrams so as to be seen from Third Dimension semicircle circle two isosceles triangles touch without their bases two rectangles with a common side and . . .

•

Fourday 28 Septober 2099

Albert's message is now clear to me, though his grammar is sometimes a bit overstretched – I suspect because he had trouble finding the exact word in his book, and settled for something close enough to be comprehensible. Most of the message is an extensive and unfortunately somewhat ambiguous series of instructions for constructing some quasi-geometric diagram. It must have some cabalistic significance, because at first glance it seems to be completely structureless.

I've got some sticks and I've dumped them on my bedroom floor. The result is my best guess at the outlines the secret message so cryptically described.

[Added later: My diagram, I now understand, looked roughly like this when viewed from the Third Dimension:

However, I had got the third symbol upside down, thanks to a misinterpretation of Al's ambiguous instructions. And he claimed to be a mathematician! In mathematics, Albert old fruit, the keyword is precision, *OK?]*

I am now wondering *where* to place my diagram so that it will be visible from the Third Dimension. I'm totally baffled – Al's message gives no clue.

[Added much *later: It hadn't then dawned on me that* every-where *– including the insides of my cupboards* and *the insides of* me *– is visible from the Third Dimension. Yes, I'd read the bits where old Albert said just that, but they hadn't registered. Idiot.]*

I'm absolutely shattered, Diary Dear, and it's time I got some beauty sleep. The rest of the puzzle must wait until tomorrow.

THE VISITATION

With so many things whirling round and round inside her head, Victoria had difficulty getting to sleep. Eventually she dozed, but her sleep was troubled with strange dreams. Somewhere in the nogonsland between 'asleep' and 'awake', with her mind drifting but half-aware, she began to hear a voice. At first it was faint – no discernible words, just a speech-like intonation, barely above the threshold of perception – but slowly it became louder and louder . . . until, with a start, she came fully to her senses.

An atavistic tingle ran along her length, and a sentence from Albert's book came unbidden into her mind: '*I became conscious of a Presence in the room, and a chilling breath thrilled my very being.*'

The Sphere! It must be somewhere nearby! Hovering just a short distance away from Flatland in that enigmatic Third Dimension! At this very moment it was experiencing a panoramic view of her house, her room, and her own internal organs. Including, she realized with shock, her dinner in the process of digestion. How yucky! Albert had never mentioned this, although he'd been surprisingly northfront about the visibility of his insides: no doubt he'd considered this particular aspect of three-dimensionality just a little *too* gross to put in his book.

The voice came again, and this time she understood its words. 'Excuse me? Someone called?' Then, suddenly, incredibly, before her eyes, a dot appeared in the middle of the room, like a very short woman except that its shading was more like an extraordinarily tiny circle.

It was orange.

It did not enter through the doorway: it was just *there*, when a moment earlier that place had been unoccupied. Was this The Sphere? She'd expected it to be larger. But then, Albert had said that The Sphere was composed of Circles of many sizes, shrinking right down to a point, so some would have to be extremely tiny.

A little way off a second dot appeared, which also expanded to a small orange Circle. Were there *two* Spheres? A double visitation?

The disembodied voice spoke again. 'Hello? Is anyone at home? Did you want something?' Then a third orange Circle, somewhat larger than the others but still smaller than a typical Circular Flatland adult, *grew* into the space between the first two:

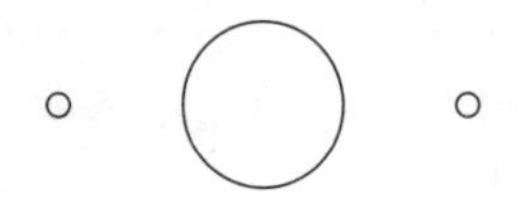

Three Spheres? How many were there? Why were they all clad in orange garments? (Were they Buddhists? Is that why they were budding?) Could there be an entire *herd* of Spheres lurking just . . . *below* (above?) the homely plane of Flatland? *What else was out there?*

It was a chilling thought, but there was no question in her mind that she must respond, quickly, while there was still a chance. If the Spheres made no contact, they might never return. Fortunately the house, being rather old, had thick walls. In a voice that she hoped could be heard by the Spheres but not by her parents, she said, 'Yes, it was me, I'm at home.' Feeling very foolish, and realizing that what she'd said didn't convey a great deal of information, she added: 'My name's Vikki Line, and I . . . I want to meet a being from the Third Dimension.'

The three Circles did something very complicated and merged together like pools of liquid, turning briefly into a shape that her sense of shading told her must be something like this:

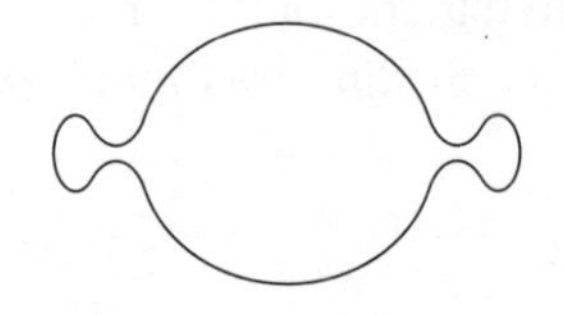

Then, so fast that she couldn't follow the movement, it expanded like a balloon into a fairly normal-looking Circle. The only difference was that the Circle didn't stay quite the same width all the time – it made her feel slightly giddy to watch it.

'Is that *all*?' said the Visitor. 'Only the *Third* Dimension?'

Vikki felt that the Third Dimension would be a good start, but Albert had waxed lyrical about even more esoteric realms, so she chose her words carefully. 'As many dimensions as you prefer, sir, as long as it's more than two.'

'Then you're in luck,' said the Visitor. 'When it comes to dimensions, I'm as versatile as the next being.' There was a pause, and he added, 'Well, more so, since *you* are the next being, and you're a Flatty.'

This didn't sound very polite, but it seldom pays to point out such social nuances to transcendent beings from another universe. 'Excuse me, sir, but are you a Sphere?'

Her stomach vibrated to the sounds of deep laughter. 'Me? A Sphere? Nothing so prosaic, I'm delighted to say.'

'Um, then . . . what are you?'

'I,' the creature declaimed proudly, 'am a Space Hopper.'

'That's a very strange name,' said Vikki, before she could stop herself.

'Not at all,' said the being, 'it's a very good name.'

'Why?'

'Because – isn't it obvious? Because I can hop spaces.'

•

Fiveday 29 Septober 2099

It took a while to sort out what the Space Hopper meant by that, Diary Darling. For a time I thought that by 'space' the creature meant 'gap' – as we would refer to the space between two buildings, say.

Not a bit of it.

To the Space Hopper, a *space* was an entire world! More than that, it was an entire self-contained UNIVERSE. Flatland is one kind of space, and Albert's freaky Spaceland is another kind of space. But that's just the tip of the iceangle. Al himself talked of Lineland, with its smartarse King, and the complacent God of Pointland: '*Infinite beatitude of existence! It is; and there is none else beside*

It.' And he dropped heavy hints of '*mysteries yet higher . . . Extra-Solids . . . and Double-Extra Solids . . .*'

Now: you and I may find this kind of stuff way out, but to the Space Hopper it was kind of boring. Extra dimensions? Yesterday's news. He – I'll call him a he, even though we haven't discussed Space Hopper Gender Issues . . . for all I know they have seven sexes or none . . . Anyway, he sounds like a male, and he's into far heavier stuff than that. One-and-a-quarter dimensions, would you believe? Infinite dimensions. Bent space and rippled space and space that blows itself up . . . Space that's just disembodied dots, space that isn't really space at all. This is one seriously spaced-out guy, ha ha.

And when he calls himself a Space Hopper, able to 'hop spaces', what he seems to mean is that he can hop BETWEEN spaces. Only there isn't any Between, just the spaces themselves. Or maybe there is a Between (I reckon he's a bit confused about this) and they all live inside some kind of Grand Unified Metaspace (or GUM) known as the Mathiverse. Look, I don't understand this buzz any more than you do, Diary Dear, so there's no use asking me for clarification. Not yet, anyway.

We ran out of chat time rather suddenly when Lee woke up and I heard her snooping outside my room. But The Space Hopper says he'll visit me again tomorrow night. And this time he'll bring a Virtual Unreality Engine.

At least, I think that's what he said.

•

Jubilee Line was getting very worried indeed. Modern morality was all very well – no, actually it *wasn't* all very well at all, it was far too lax, though she acknowledged with a certain wistfulness that she had been born a generation too soon – but Grosvenor would have a fit if he found out that his dear sweet Victoria was entertaining a man in her room in the early hours of the morning.

There was no doubt in Jubilee's mind. She had woken up, and heard muffled voices coming from somewhere in the house. At first she'd thought it might be burglars, and she'd sneaked out of her room with every intention of stabbing the intruder with her sharp rear end. She had neither fears nor qualms about this: all Flatland women were armed and dangerous by virtue of their needle-sharp lineal geometry, and they were trained from birth to use this natural advantage if it became necessary. Criminals had no rights in Flatland – rights were what you got if you agreed to abide by the common legal code, and you couldn't have things both ways. But as she sneaked closer to the source of the sound, it resolved itself into a conversation between an unknown male (she could tell by the voice) and her daughter! Her worst fears were being confirmed; she'd known Vikki's change of habits had to be a symptom of some kind of improper behaviour.

It was that computer, without a doubt. She never should have let Grosvenor buy the thing. Not that there'd been any way to stop him once he'd got the idea into his head, however damnfool it was.

She tiptipped closer. Suddenly the voices stopped, but not before she heard Vikki agree to meet the stranger in her room the following night. *Well, we'll see about that!*

As quietly as she could, Jubilee made her way back to her own room. Silence once more descended on the Square household. But now it was a *thoughtful* silence.

•

Vikki *had* heard right. The thing that the Space Hopper was going to bring was indeed a virtual unreality engine, or VUE. It would, he said, 'blow her mind'. Not literally – despite its ungainly appearance, the gadget was perfectly safe, because most of it existed only in the metareality of the Mathiverse. More precisely, the Space Hopper explained, the VUE would expand her mental horizons. Take her mind to new places. Change her point of VUE. She was about to become a Space Traveller, and the VUE was to be her Space Suit. With its extradimensional terminals plugged into the delicate tissues of her brain – an easy operation, since her brain was protected only against interference from North, South, East, and West, not from Up or Down – she would be able to experience each of the bizarre spaces of the Mathiverse *from the inside.* The Space Hopper had no need of such prosthetics: he could transport himself from one space to another using only the Power of Imagination. A Flatland mind, though, would need technological assistance to overcome its own inherent limitations.

So absorbed had Vikki and the Space Hopper become in installing the drivers for the VUE and customizing her user-preferences that they failed to hear Jubilee's approach. Through the solid door, Lee could hear their discussion clearly. Whatever they were doing, phrases like 'reinsert your floppy' sounded distinctly immodest and unladylike.

When the strange male urged Vikki that it was time to 'go for it', Jubilee could no longer restrain herself. Outraged, she threw open Vikki's door, and charged into the room, her infinitely thin needle-point waving threateningly. Ahead was her daughter, frozen by the suddenness of the intrusion . . . and a remarkably handsome Circle wearing the most ghastly orange cloak.

Jubilee skated to a halt, confused. At least her daughter had *taste* – more than she could say about the man's dress sense. Jubilee wondered how Vikki could have met this gorgeous hunk. But no good ever came of affairs with the upper classes – they'd just get you segment and dump you. With a shrill cry of 'You swine! You ought to be circumscribed!' Jubilee launched herself backwards at the intruder, intending to give him a brand new diameter. But he just seemed to slide away from her infinitely sharp tip. She *thought* she heard Vikki's voice saying 'Don't worry, Mother . . .' Then, silence.

Vikki, and her high-bred paramour, were gone.

With growing horror, Jubilee searched every corner of the room. She *knew* they hadn't gone out by the door, and that was the only possible route. They'd vanished.

Eventually, Grosvenor woke up to find his distraught wife peering hopelessly behind every piece of furniture and every child-hood toy in their daughter's room, sobbing her midpoint out.

•

Virtual Unreality, Vikki had decided, was *cool*. Agreed, it took a bit of getting used to – but once you'd mastered the knack of seeing new geometries in their own terms, instead of expecting them to have the same properties as the space you were familiar with from birth, you could begin to appreciate their novelty. And it certainly made you realize just how limited your familiar concept of space was. Having the VUE made all this a lot easier: without it, you'd have to rely on your imagination. In Virtual Unreality you could *experience* what others could only dream about. She now understood how Albert had felt when his flat world had suddenly expanded into the glorious realm of solidity, literally opening up new directions. And the true form of the Space Hopper was now visible to her, with his twin horns, manic grin, and bulging body. He wasn't one circle, or two, or three. All the while she had been trapped within the confines of Flatland, all she had been able to see of the Space Hopper was a two-dimensional cross-section. As he moved up or down (words to which she could finally put a meaning), his cross-section changed. As he rose through the plane of Flatland, what she had seen was this:

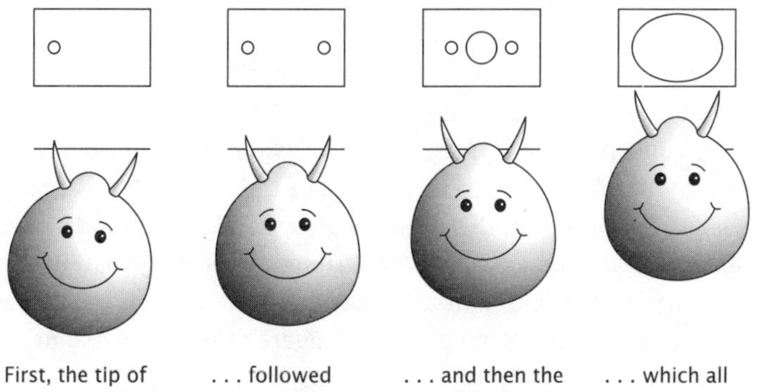

First, the tip of one horn poked through . . . | . . . followed by the tip of another . . . | . . . and then the top of his head too . . . | . . . which all merged to become his body

Even the complicated shape she thought she had seen, if only for a moment, now made sense: it was a slice through the level where the horns joined the bulbous lump of the creature's body.

So *this* was what Albert had seen! Not with a shape as complicated as the Space Hopper, though, just a straightforward Sphere:

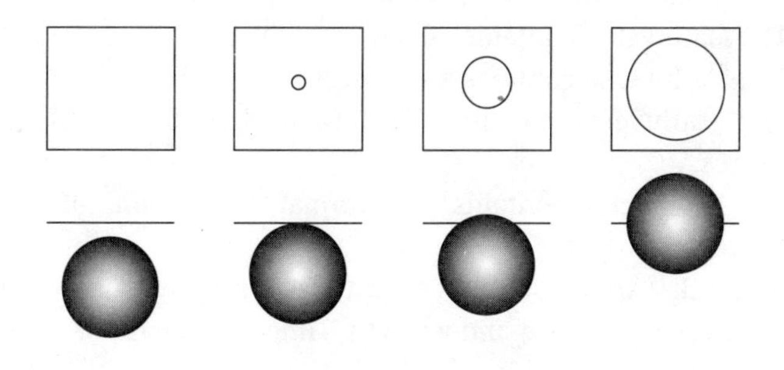

As the Sphere rose towards the plane of Flatland, not yet intersecting that plane, it was invisible. As its surface began to pass through that plane, it appeared – from nowhere, it seemed – as a single point, which became a small circle, then a larger one, then . . . When the 'equator' of the Sphere aligned itself with the plane of Flatland, the circular cross-section had the same diameter as the Sphere itself – but what a woefully inadequate representation this single circular slice was! And if the Sphere continued to rise, its cross-section would shrink back down to a point, and then vanish.

•

What is the Mathiverse?

The Mathiverse transcends Time and Space . . . it transcends Intelligence and Extelligence . . . it transcends Thought; it transcends Transcendence itself. Within it – and 'within' is definitely the wrong word, for concepts such as 'inside' and 'outside' apply to individual Spaces, not to the unfathomable reaches of the Mathiverse – are not just all Spaces and Times that *have* existed, or all Spaces and Times that *will* exist, or even all Spaces and Times that *could* exist. It also contains (wrong word, again) all Spaces and Times that could *not* exist, if only as a grim warning of the dangers of the nonexistent.

The Mathiverse contains all numbers.

The Mathiverse contains all shapes.

The Mathiverse contains all geometries.

The Mathiverse contains all vectors, matrices, permutations, combinations, integrations, separations, projections, injections, surjections, bijections, semigroups, transformations, relations, functions, functors, functionals, algebraic group schemes, supermanifolds, *K*-theories, *M*-theories, *M*-sets, power sets, subsets, supersets, and plain, ordinary, common-or-garden sets.

The Mathiverse contains all data structures.

The Mathiverse contains all processes.

The Mathiverse contains all formal descriptions of logical structures.

The Mathiverse contains all informal descriptions of illogical structures.

If one day somebody managed to invent a new kind of *thing*, one that wasn't a Space and wasn't a Time but somehow belonged in the same category (and, now that you mention it, the Mathiverse contains all categories) . . . anyway, if somebody did what I've just said, then whatever they came up with would have been present in the Mathiverse all along. (Except, as you've guessed, 'would', 'have', 'been', 'present', and 'in' are the wrong words, and so are 'all' and 'along'. We can probably accept 'the', though.)

Space Hoppers use the name 'Planiturth' to refer to the world that you, dear reader, inhabit. Planiturth is *not* the same as what old Albert had called 'Spaceland' – but there is a close connection. Spaceland is an abstraction that captures some of the essence of Planiturthian geometry, but Planiturth is *real*. Planiturth has a history as well as a geography, and its relation to the Mathiverse is complex and convoluted.

Even philosophers (a special breed of the creatures that live on Planiturth, generically known as Peoples) do not doubt that Planiturth is intimately associated with the Mathiverse – *of* it but not *in* it, so to speak (not that 'in' . . . but we've been through that already, sorry). In a sense, the Mathiverse is a creation of the combined mentality of the Intelligences of Planiturth. It is a Planiturthian mental construct. Despite which, it has its own kind of reality – it is *so* real that every atom of Planiturth's universe dances to its tune. The very rules by which Planiturth's universe runs are drawn from the Mathiverse.

Or so the Planiturthians think, anyway.

It is a deep philosophical conundrum. Does Planiturth's universe *really* obey rules that come from the realms of the Mathiverse? Or is that an illusion born of Planiturthian prejudices? Is Planiturth's universe built from *mathematics*? Or is mathematics built by the minds of Planiturthians? Planiturthian mathematicians would like to think that their universe is built from mathematics, but that's only natural, after all. Planiturthian physicists would like to think that the Planiturthian universe is built from physics. Planiturthian biologists would like to think that the Planiturthian universe is built from biology. Planiturthian philosophers would like to think that the Planiturthian universe is built from philosophy. (Let me tell you a secret: it is. The fundamental unit of the Planiturthian universe is the *philosophon*, a unit of logic so tiny that only a philosopher could hope to split it.) Planiturthian greengrocers would like to think that the Planiturthian universe is built from carrots and potatoes. Planiturthian taxi-drivers have the definite opinion that the Planiturthian universe is built from opinions, and they are prepared to say so at length. And Planiturthian beauticians would like to think that the Planiturthian universe is built from beauty, and in this belief they may be closer than all the others, for whatever attributes the Mathiverse might have, beauty is paramount among them.

Mathiversian beauty is, of course, an acquired taste.

The Planiturthians are unique among the creatures of (but perhaps not *in*) the Mathiverse in that they are not sure which Space (or Time, or Spacetime, or . . .) they really belong to. Long ago they thought they inhabited Spaceland, a rigid 3D universe; Time was simply something that passed, a separate 1D process. This was the Clockwork Universe of a famous People named Isaacnewton: Absolute and Uniform Space ticking to the beat of an Absolute and Uniform clock, and decorated with Absolute (but definitely *not* Uniform) Matter. Spaceland is a genuine part of the Mathiverse, but the Planiturthians have now decided – correctly – that they don't actually live in Spaceland at all. They live, perhaps, in Curvyspace. Or is it in Waveworld? Or are they gambolling in the Quantum Fields ('gambling' might be a better word since these fields are made from *probabilities*). The Planiturthians aren't really sure, and about every ten years they completely change their minds.

Mind you, Flatlanders will shortly have some rethinking to do, too.

•

Sixday 30 Septober 2099

We've been exploring Spaceland! (Planiturth, too, but that's a lot more complicated.) Like Flatland, it's a *Euclidean* space (as I now know to spell it). When the Space Hopper first said that word, I asked him what a 'yuke' was and why it needed a lid. He hummed and hawed for a while, and then said that somewhere in the humongitude of the Mathiverse a Planiturthian being known as Euclidthegreek first mapped out the geometry of worlds like Flatland.

I find this a questionable assertion, to say the least, since according to all the history books Flatland was first mapped out by an explorer named Marco Polygo. Still, I rely on the Space Hopper to conduct me from space to space safely, and it is best not to annoy him.

Anyway, Diary Dear, it's amazing what an extra dimension can do for you. Not just Spheres and Cubes and Pyramids and things – face it, those are pretty obvious extensions of Circles and Squares and Triangles. A Flatlander, shown a sequence of cross-sections, could get quite a fair idea of what they're like. But there are MUCH more interesting things in Spaceland, things that are completely outside the Flatland frame.

Some of them are simple. In Flatland, if you hammer a nail right through a plank of wood, it falls to bits:

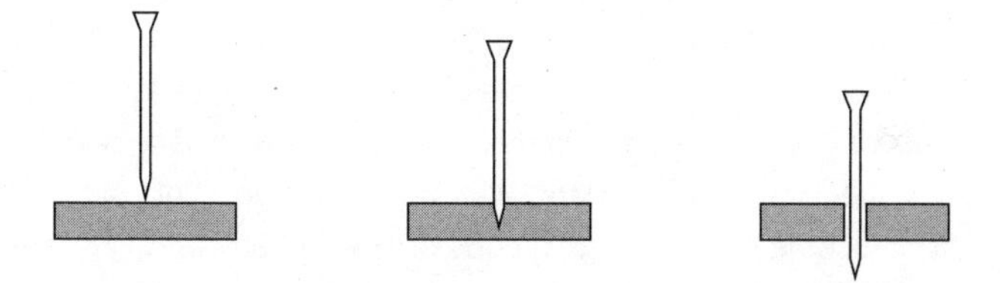

This DOESN'T happen in Spaceland. And they can use *wires* to connect telephones, without trapping everyone in the tangle they create. They put the wires UNDER (a new word I've learned that means 'in the Down direction') the places where people walk.

Some are subtle. My absolute FAVE is a widget they call a KNOT. This is a loop of string, like a Circle, but it sort of runs through its own inside!!! The result is that it tangles up in a way you can't rearrange into an ordinary Circle. Flatland is just too low-dimensional to allow knots.

The Space Hopper says that even Spacelanders (well, actually he called them 'Planiturthians', he says there's a subtle distinction but he refused to talk about it until later) don't actually have a very good feel for 3D – they have trouble with *knots*, for instance! This, he says, is because their eyes really see only in 2D. They see a 2D *projection* of an object, not the entire object ALL AT ONCE. So it's surprisingly easy to shake their intuition about the space they live in.

For instance, here's a teaser that the Space Hopper asked me. Take a Spaceland cube, of unit side (that's a fancy way to say 'one thingy each way' where the 'thingy' is some standard unit of measurement, OK?) and cut a hole through it (without it falling apart, like a plank would in Flatland) with a square cross-section, so that another cube can be pushed through the hole. Got that? Right – here's the question. What's the largest size of cube that can be made to pass through such a hole?

I know what you're thinking, Diary Dear. Obviously you can't push a larger cube through a smaller one. But that's a Flatland way of thinking. What you have to realize is that the second cube might be pushed through at a slant. With the *right* slant, you can push a cube of side 1.06 through a cube of side 1. You cut it like this:

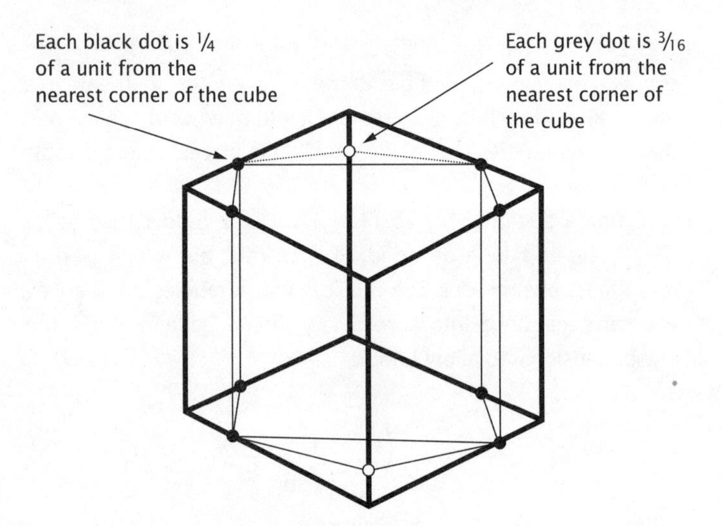

So in Spaceland, you can push a bigger cube through a smaller one! (Not many people know that.)

On the other hand, there are *some* things that Spacelanders THINK they know – and it turns out they're RIGHT – but it's not so easy to PROVE they're right. One of the most amazing is all about how greengrocers stack oranges . . .

•

'It's called the Kepler problem, and it's one of the oldest unsolved Planiturthian mathematical problems,' said the Space Hopper. 'At least, it *was* until someone solved it. Then it was one of the newest *solved* Planiturthian mathematical problems.'

'How old was it? Before some Planiturthian solved it, I mean?'

The Space Hopper scratched one of his horns. 'In the Planiturthian calendar it goes back to the year 1611. That's 1711 in Flatland dates – by a strange coincidence, Flatland time runs exactly 100 years ahead of Planiturthian time. And it was solved in 1999 their time, so it lasted 388 years!'

'What's it about?'

'The best way to stack spheres. Tell me, Vikki, how do greengrocers stack fruit in Flatland? Round fruit like . . . well, roundfruit?'

That was an easy one. As a small child, Vikki had always enjoyed a visit to the greengrocer's hexagonal shop with its neatly laid out clumps of fruit. The squarefruit, of course, were always packed in squares, because that way they fitted exactly. But the roundfruit were packed in a honeycomb pattern. That left gaps, but you had to leave gaps, and in this arrangement the gaps were as small as they could be:

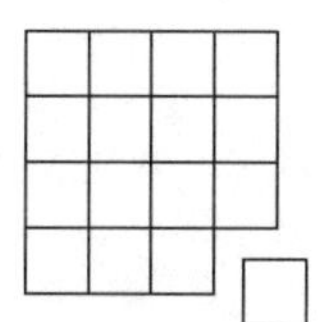

Squarefruit fit together tightly like this

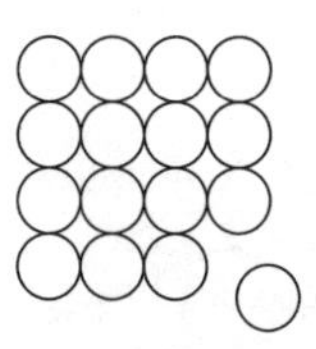

You **could** pack roundfruit like this . . .

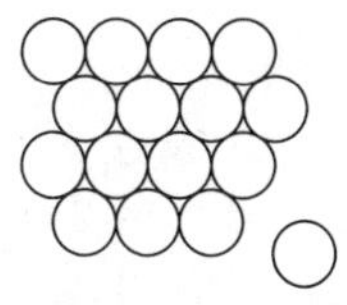

. . . but **this** way you get more of them in a given space

Flatlanders liked both arrangements, because the first was based on squares and the second on hexagons. (If you can't see the hexagons, look at any circle and the circles that surround it. There are six of them, and their centres form a regular hexagon.)

'Quite right,' said the Space Hopper when she told him this. 'But how do you *know* that the hexagonal pattern is the most efficient?'

'Experiment?' suggested Vikki.

The Space Hopper shook his horns. 'No, that's not a proof. It's one of those cases where the answer is "obvious", but proving the obvious is difficult. It took the Planiturthians until 1910 to do that – and even then the proof outlined by Axelthue is rather sketchy. A really solid proof didn't turn up until Fejestóth found one in 1940. So don't be too unhappy that you couldn't tell me why the hexagonal pattern is the most efficient way to pack circles in the plane. It's hard, however obvious it may seem.

'Spacelanders have fruit too. They have bananas, which are long thin ellipsoids – usually bent – and they have spherical ones, like oranges. And, just as in Flatland, every Planiturthian greengrocer *knows* the best way to pack oranges. How do you think they do it?

This was a good test of Vikki's mastery of the VUE, and she created lots of virtual spheres in mid-air and started fiddling with them. Eventually she stopped. 'How about this?'

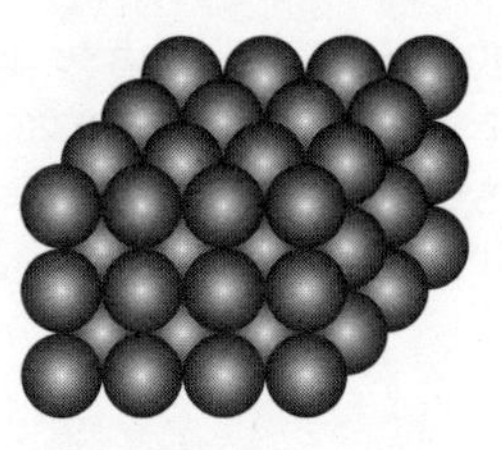

You **could** pack oranges like this . . .

'That would be good with cubefruit,' said the Space Hopper. 'But even in Flatland, the way you pack squarefruit isn't the way you pack roundfruit. You've arranged each layer in squares. Try again.'

'Um . . . Oh, I see! I can make lots of hexagonal layers and stack them vertically on top of one another!'

. . . but **this** way you get more of them in a given space . . .

'Better – but you're still thinking like a Flatlander. You can do better still.'

'I don't see how.'

'Try shifting the layers so that the bumps in one layer fit into the dips in the previous one. You're putting bumps against bumps.'

Vikki thought about this for a moment, then she rearranged her virtual oranges into a more efficient packing:

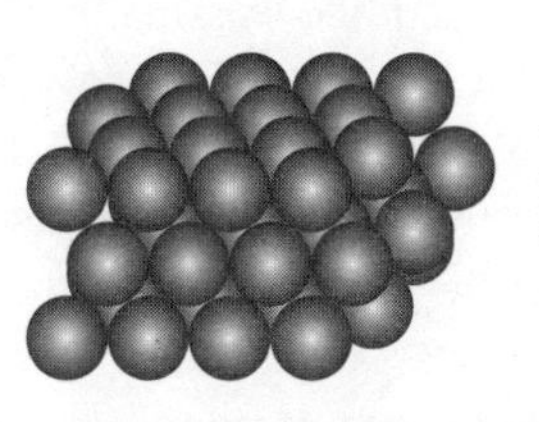

. . . and **this** way you get even more of them!

'Fantastic!' said the Space Hopper. 'That's the one I want! As I mentioned, it's the arrangement that every Planiturthian greengrocer knows fills space most efficiently. A Planiturthian mathematician and mystic called Johanneskepler knew it too, in 1611 – or, like the greengrocers, he thought he did. It's certainly a very efficient way to pack spheres in 3D. But is it the *best* way? That's another question altogether.'

'Does it really matter? After all, if I'm a greengrocer, I don't really care as long as I don't waste too much space.'

'And if you're a Planiturthian, as long as the pile doesn't collapse. Quite right – and that's one reason why the greengrocers *don't* "know" the answer. What they know is the answer to a slightly different question. They build their piles of oranges on a flat surface, and mostly they want to find a method that avoids avalanches. But in a way it *did* matter to Johanneskepler that this arrangement is the best, though he didn't realize it at the time. What he wanted to understand was snowflakes.'

'Surely he didn't want to stack snowflakes?' asked Vikki, puzzled.

'No. It's easy to stack snowflakes: it's called a snowdrift. Though that doesn't stack them very efficiently. No, Johanneskepler wanted to understand why snowflakes are six-sided. You see, he wasn't very rich (being a mathematician), and he needed to give his sponsor – who rejoiced in the name of Johnmatthewwackerofwackenfels –

a New Year's present. So he wrote him a book called *On The Six-Cornered Snowflake*, in which he tried to work out why snowflakes had sixfold symmetry. And he traced it to a packing problem.'

'But you just said—'

'Ah. Not a problem about packing snowflakes together. A problem about packing together whatever tiny units of *stuff* a snowflake is made from. Johanneskepler didn't have the modern concepts, but what he was asking about was the atomic structure of ice crystals. He figured out that *if* ice was made of tiny spheres – not quite right, but close enough – and *if* those spheres packed together as closely as possible, then the "crystal lattice" of ice would have a hexagonal structure. Just like the hexagonal packing arrangement for oranges. And the second "if" was spot on, because atoms pack together in whatever arrangement makes their energy lowest, and basically that means close-packing.'

Vikki knew that 'atoms' were very small particles – well, particle-like things – out of which all matter was made: they'd done this in Physics. And she knew about crystals. They were made from atoms arranged in a regular lattice pattern, a grid. In this respect, Flatland was just like Spaceland, except that there were 17 types of crystal lattice in Flatland, as compared with 230 in Spaceland. So what the Space Hopper was telling her all made perfect sense – for once.

'Anyway,' the Space Hopper went on, 'there was a period of nearly four hundred years when the Kepler conjecture – that the arrangement of staggered hexagonal layers is the best way to pack spheres – was something that "most mathematicians believed, and all physicists knew". But you can imagine that if a similar problem in 2D wasn't solved until 1910, then its 3D analogue would prove to be a lot harder – and it was. Lots of Peoples tried, and failed. Then, in 1999, a Planiturthian mathematician named Thomashales finally proved that the physicists had been right all along. (This greatly disappointed some of the mathematicians, who were vaguely hoping that the physicists would turn out to have been wrong. Fat chance!) His proof was *huge*: he turned the problem into a vast number of calculations and then got a Planiturthian computer to do the sums. The idea is to start with an arrangement that's *not* packed into staggered hexagonal layers, and show that by rearranging some of the oranges you can improve the efficiency.

Because in principle there are many possible ways not to have staggered hexagonal layers, you have to do this in an awful lot of cases. But Thomashales managed to organize a *huge* calculation to deal with every possibility.'

•

Sevenday 1 Noctember 2099

Planiturth, and its Spaceland geometry, are BRILL. I'm having my mind stretched in a dozen directions at once, and they're all fascinating. But tonight I'm feeling a bit sad – homesick, really. My parents must be terribly worried. Thing is, I'm having a lot of fun, and I don't want to go home JUST yet – but it would be nice to get some kind of message back to them to say I'm OK.

Maybe there's a way to tap into the Flatland phone system from here. I'll ask the Space Hopper tomorrow.

4

A HUNDRED AND ONE DIMENSIONS

Wunday 2 Noctember 2099
I've been MUCH too busy to phone home.

Compared with Flatland, Diary Dear, the Planiturthian universe is a riot! For starters, it's got a HUMONGOUS number of Spheres – well, they're more or less that shape – which range from huge balls of fire called *stars* to tiny balls of ice known as *setalights*. No, Diary Darling, I haven't the foggiest *why* they call them that – I guess because you have to set them alight to warm them up. (It'd be a better name for the stars, really.) Anyway, the Space Hopper has been having a whale of a time buzzing all over the place, showing me all the tourist attractions like Black Holes and Ooooooooort (I *think* that's the spelling) Clouds . . .

I'd call it a whistle-stop tour, except we never *stop*. Or whistle.

The Planiturthian universe, I now know, contains a vast range of shapes that I've never seen before. Some are GIGANTIC: spiral things called galaxies, clusters of galaxies, superclusters . . . right up to rippled skeins of matter millions of light years across. (A light year sounds like a unit of time but it isn't: it's a very big unit of distance indeed, the distance light goes in a year. And in the Planiturthian universe, light *sizzles*.)

On a smaller scale, most of the Planiturthian universe seems to be made of rocks. Of every conceivable shape and size. And many inconceivable shapes and sizes. Except – most of the Planiturthian universe is uninhabited, but every so often you find *living creatures* of one kind or another. Even intelligent beings like US (though not as beautiful and *far* less perfect). Almost without exception the inhabitants of the Planiturthian universe (oh, yes, there's lots) are convinced that they are the ONLY lifeforms in the whole of Space. The Space Hopper says that when it takes light several centuries to get to your nearest neighbour, then it's hardly surprising that you should think you're the only lifeform in the whole of Space. I think the Space Hopper is much too willing to tolerate poor imaginations, but that's just the opinion of a humble Flatlander.

One of the smaller spheres (it would be tempting to call it Roundworld, but there are SO MANY round worlds here, and anyway its inhabitants refer to it as Planiturth, so I shall too) is quite a fun place. I've told you about it before, but now I've had a crash course in Planiturthian Studies. Lots of these medium-sized Spheres are littered with common – though distinctly nongeometric – shapes, such as rocks, mountains, canyons, volcanoes, and clouds. Planiturth is smothered from pole to pole with far more unusual structures: swamps, trees, plants, animals, and curious vaguely star-shaped things with five protuberances that call themselves *Peoples*. Like this:

The Peoples of Planiturth are just as complacent about living in a 3D world as we Flatlanders are about living in a 2D one. Most of them are convinced that there is no such thing as the Fourth Dimension. The Space Hopper says that they're right, but for the wrong reason.

I wonder what he means by that? Tomorrow, I'll ask him. Right now, I'm worn out. I do wish I could phone home. Maybe I can build a transmitter!

•

'They're right, Vikki, because there is no such thing as *the* Fourth Dimension. And they're wrong, because there are lots of different Fourth Dimensions – not to mention Fifth, Sixth, or even a Hundred-and-First – many of which they experience in their daily lives, but fail to recognize.'

Vikki found it hard to believe that anyone could experience a Fourth or Fifth Dimension and not know it, and said so.

'It's a question of being sensitive to what you're actually experiencing,' explained the Space Hopper. 'They've got it into their heads that "Fourth Dimension" has to mean a fourth direction in *space* – by which they mean their usual 3D space. Well, naturally if you go looking for a Fourth Dimension in a 3D space you won't find it, any more than you could find a Third Dimension in Flat . . . *Well*, actually, that's not quite—'

'You're saying they should look for a Fourth Dimension *outside* their own space? *You* may be able to hop spaces, but for the rest of us it's not so easy, you know.'

The Space Hopper wiggled impatiently. 'Sometimes. Not always. It may be just a question of looking at their own space in a different way. But instead of us just talking about it, let's see what a People has to say about it.' He made some adjustment to Vikki's VUE, and the universe turned inside out . . .

An infinite series of folds unfolded, and Space stabilized. Vikki sat still for a moment until she stopped feeling sick.

They were sitting at the side of a rutted dirt path in the middle of beautiful, rolling countryside. Beyond it was a field full of fluffy white things on sticks that made 'baaaa' noises and copied each other a lot. From the distance came a curious sound, a kind of rattle-buzz-hum. Then, along the path, bouncing erratically from one rut to another, came A Machine. It rolled along on two big Circles, and a People sat precariously on top of it.

'A Bicirclist,' whispered the Space Hopper. 'We're very fortunate: you don't often see them in this habitat.' He bounced up and down, and the Bicirclist promptly left the path and toppled into a ditch. There was a silence, followed by a stream of Planiturthian invective. The Bicirclist climbed back onto the road and stomped towards them.

'*Why did you*— Oh, I say, are you two part of some promo or something? How on Earth do you squeeze into those costumes? Are you dwarves? That squarey thing is amazing, it looks so thin that from the edge it seems to vanish altogether! Some kind of optical illusion? Must be.'

'Not at all,' said the Space Hopper. 'We are visitors from a different Dimension.'

The Bicirclist clutched its belly and laughed heartily. 'I know,' it said, 'you're students, aren't you? Is it Rag Week? This is some kind of practical joke.'

The Space Hopper turned to Vikki. 'You see?' She nodded.

'Why do you think it's a joke?' asked the Space Hopper.

'Because there aren't any Extra Dimensions,' said the Bicirclist. 'There's the usual three, and that's it.'

'Really?' said the Space Hopper. 'Just suppose, for instance, that your comfortable, solid, 3D space were in reality just an infinitely thin slice of a far more glorious 4D continuum.'

'*Slice?*' said the Bicirclist, incredulously. 'How can a solid space be a slice?'

'Let me tell you about a place called Flatland . . .' the Space Hopper began.

•

The Bicirclist was extremely patient.

In the circumstances.

•

'So you see, a Flatlander must face much the same question: since the plane is all there is, how can it possibly be a slice of *anything*?' For the dozenth time, the Space Hopper had launched into his favourite argument. 'By analogy, in what sense is a plane *thin*, when it extends infinitely far along both of the possible dimensions North–South and East–West? There is no other direction for it to be thin *along*. It's thin along the Up–Down direction? These words have no meaning. Show me "Up" and I might believe you – but you can't show me "Up," not if we remain in Flatland. In the same way, you Spacelanders may have difficulty comprehending that your solid 3D space might indeed be an infinitely thin slice of something bigger.'

'But that's just an analogy.' The Bicirclist was getting really into his stride now, and his face was flushed and animated. He waved his protuberances a lot. 'The difference is that I can *see* there's more to Space than just a plane. But I can't *see* there's more to Space than . . . Space.' Before the Space Hopper could tell him yet again that this was because his perceptual apparatus was too limited, the Bicirclist quickly conceded part of the argument. 'I suppose it might be a part of something bigger that we can't perceive,' he said, 'I grant you that. But whatever it is, it can't be *thin*.'

'Oh, but it can', the Space Hopper contradicted. 'Agreed, your space is very thick (though probably not infinitely so) along the three dimensions of North–South, East–West, and Up–Down. But there could be a fourth dimension, Chalk–Cheese, and along that direction your supposedly thick space might be no more than a filmy slice. Though the "film" would be three-dimensional, of course. Travel a trillionth of a micrometre Chalkwise or Cheesewise, and you would move out of your homely Space altogether!' And he scratched a rough diagram in the dried mud at the side of the path:

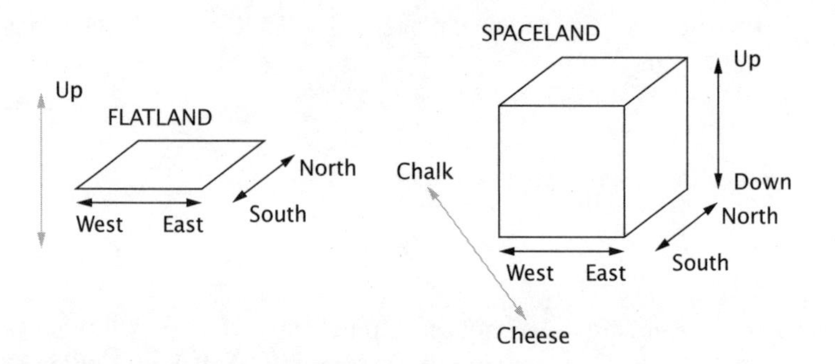

'It's not very convincing,' said the Bicirclist. 'I don't quite see where the Chalk–Cheese line *goes*.'

'Yes, well . . .' The Space Hopper writhed with embarrassment. 'I . . . yes, I admit it's not possible to draw a convincing fourth dimension on a two-dimensional path – but then, even to represent a *third* dimension requires the convention that a slanting line doesn't actually reside within the picture.'

'I guess.'

'So we just need a further convention, that another kind of slanting line doesn't reside within 3D Space.'

'Mmmph.'

The Space Hopper could see that the Bicirclist was still unconvinced. 'OK, then: replace "Chalk" with "Past" and "Cheese" with "Future".'

'I don't catch your drift, old chum.'

'Isn't your present world separated by a trillionth of a second from its future, and likewise from its past? *Time* is an example – but not the only one – of a Fourth Dimension. And along the time direction, your present world is wafer-thin!'

The Bicirclist disputed that time could be a dimension, but as usual all he really did was argue that it wasn't a spatial dimension, which was reasonable but irrelevant. The Space Hopper turned to Vikki. 'Wouldn't it be nice if one day a Hypersphere could convince him of the existence of a Fourth Dimension, merely by appearing from thin air within a locked room, growing from a point to a small sphere to a larger sphere, and then shrinking back to nothingness.'

'Or maybe he could get a visitation from a Hyperspace Hopper from the Fourth Dimension!' said Vikki. She paused. 'Of course, all he'd see would be a puzzling collection of spheres, changing size before his eyes.'

'Exactly. Only a Virtual Unreality Engine could reveal the extradimensional logic behind the strange contortions.'

•

In the wake of Victoria's baffling departure, the Square household had become a very gloomy place indeed. In Grosvenor's mind, the trouble caused by his ancestor's old book and his daughter's inexplicable departure had to be connected somehow. He felt guilty that his hasty burning of the ancient document was what had upset Vikki and caused her disappearance, and to avoid admitting this – to himself or anyone else – he alternated between raining curses on old Albert for bringing the family into disrepute from the far side of the grave, and denouncing Vikki for the same offence from its near side. These episodes, in turn, were punctuated by long periods of stony-faced silence on the whole issue – which, he would declare, was just something they should put behind them. 'The girl was no good, Jubilee,' he would say, 'always obstinate and headstrong. I don't know how she vanished, and I don't know where she went, and I don't *care*! The family must move on.'

The boys, Les and Berkeley, had been told that their sister had been offered a really good job in Numerica, a distant region of Flatland, and had been forced to leave abruptly before it was offered to someone else. They appeared to accept this explanation, but Jubilee could tell they were only half-convinced.

She herself had gradually come to a very different view from the one that her husband professed to hold. She was puzzled by the manner of Vikki's sudden disappearance, and the more she thought about it, the less certain she was that her daughter had, in fact, been involved in a secret liaison. She had searched Vikki's room and found nothing incriminating; she was starting to wonder if her daughter had been an innocent victim of events that neither she nor Grosvenor understood. And she couldn't help sensing echoes of Albert's strange visitations. Just what *had* been in that old book?

Grosvenor had notified the police, of course, but he'd had to conceal some of the story because officialdom would have thought he was nuts if he'd told them she'd vanished into thin air. So he just said she'd run away. Now, a daughter of Vikki's age disappearing in a huff was hardly unusual, and technically Vikki was an adult, responsible for her own actions and her own decisions. In the absence of any signs of foul play, the police were privately regarding her disappearance as a typical flight from the family home by a young woman in search of personal liberty. They put Vikki on their missing persons list, and that was pretty much all they did.

Jubilee knew that her husband's real feelings were very different from those he dared to express, and she hoped he would eventually be able to admit as much. She was worried that when Vikki turned up again – and she had an exceptionally strong hunch that it *would* be 'when', not 'if' – then Grosvenor would blow his northedge and cause an irreparable rift.

Secretly, Jubilee also hoped that her daughter was having fun – otherwise she had caused an enormous amount of trouble for nothing – and when everything started to overwhelm her, she tried to lose herself in housework and looking after the boys.

It was a very difficult time for them all, and the atmosphere in the spacious old house was strained and awkward. Still, they coped.

•

'If Time were a dimension,' said the Bicirclist, 'you ought to be able to travel *along* it, like you can with the dimensions of Space.'

'You can,' said the Space Hopper. 'You can travel forwards at one year per year.'

'No, I mean travel back and forth at will.'

'Ah.' The Space Hopper turned to Vikki and whispered, 'Should I reveal that we Space Hoppers are also Time Hoppers?' He turned back to the Bicirclist. 'Travelling through Time is more difficult than travelling through Space because the temporal metric is far more compressed—'

But the Bicirclist wasn't listening. 'Time travel – like that bloke in the story by H.G. Wells. Which, by sheer coincidence and narrative imperative, I've got in my saddlebag.' He rummaged in a small bag attached to the rear of the Bicircle and extracted the book triumphantly. '*The Time Machine*, that's what it's called.' He opened the book and thumbed through the first few pages. 'Wells's Time Traveller opens the story by arguing that an *instantaneous* solid body is just as much a mathematical fiction as a line or a plane: "Clearly," the Time Traveller proceeded, "any real body must have extension in four directions: it must has Length, Breadth, Thickness, and – Duration". Having argued that material bodies are really four-dimensional, Wells's Time Traveller then springs his trap. "There is no difference between Time and any of the three dimensions of Space except that our consciousness moves along it," he says, and drops dark hints that he has overcome this particular limitation.'

The Space Hopper had a good grasp of Planiturthian history, especially when it came to Space and Time. 'Do you know that within ten years of *The Time Machine*, a People called Alberteinstein invented the Theory of Relativity? And another named Hermann-minkowski formulated Alberteinstein's ideas in terms of the geometry of four-dimensional spacetime?'

'Nope.'

'Well – the idea of time as *the* fourth dimension came into vogue, and in this world it's never quite gone away again. But the truth is far stranger. Your world has not just four dimensions, but five, fifty, a million, or even an infinity of them! And none of them need be time. Space of a hundred and one dimensions is just as real as a space of three dimensions.'

'Rubbish!'

'Not at all. Why, take that Bicircle of yours. It has at least seven dimensions.'

'Really? Then show me some.'

'I will if you bring it over and hold it upright.' The Bicirclist obliged. The Space Hopper bobbed across to the Bicircle. 'You'll grant that its current shape is three-dimensional?'

'Of course.'

The Space Hopper nudged the Bicircle's handlebars through a small angle. 'There! A Fourth Dimension!'

The Bicirclist laughed. 'Don't be silly! You just moved it in 3D.'

'Yes, but it changed shape.'

'Nonsense!'

'No, it really did,' said Vikki. 'If it had stayed the same shape you could make it fit into its current position by a rigid motion. But that bit with horns on had to turn while the rest didn't. *That's* not a rigid motion!'

'Oh. I suppose you could put it like that. But if it moved through a new Dimension, in what direction does that Dimension point?'

'In the Turn-The-Handlebars direction,' said the Space Hopper, 'which is different from the Turn-The-Front-Circle direction, which is different from the Turn-The-Back-Circle direction, which is different from the Turn-The-Pedals direction. That's four new dimensions, in addition to the original three. You've got a seven-dimensional bicircle.'

•

Twoday 3 Noctember 2099

I really must phone home soon. But the Space Hopper says that this will be difficult, because of system incompatibilities. Apparently, if I want to phone home then it's best not to do it from here.

Well, Diary Dear, eventually the Space Hopper managed to explain in what sense the Planiturthian People's Bicircle is Seven-Dimensional. It's all to do with *variables* – quantities that can change. 'Dimension' is a geometric way of referring to a variable. Time is a nonspatial variable, so it provides *a* fourth dimension, but the same goes for temperature, wind-speed, or the number of termites in Tangentia. The position of a point in three-dimensional space depends on three variables – its distances East, North, and Upwards relative to some reference point. By analogy, anything that depends on four variables lives in a four-dimensional space, and anything that depends on 101 variables lives in a 101-dimensional space.

In fact, ANY complex system is multidimensional. The weather in a typical Flatland back garden depends on temperature, humidity, two components of wind velocity, barometric pressure – that's five dimensions already! I didn't know we had a 5D garden before! An economy with a million different commodities, each having its own price, lives in a MILLION-DIMENSIONAL space!!

No wonder economies are hard to control!

•

Vikki was dreaming . . .

She dreamed she was floating in the darkness above a world that looked much like Flatland must have been before Life evolved – a featureless, uniform, white surface. But *something* lived on the surface, for she could hear it muttering to itself: 'Painting the dot, painting the dot, red paint, red paint, round and round; painting the dot, painting the dot, red paint, red paint, round and round . . .' over and over again.

In her dream, she came closer, and discovered a curious little antlike creature, with pot and paintbrush in hand, going round and round in a spiral. It had started by painting a small dot, and then it had gone round the edge making it a bit bigger, and it had carried on until by now the dot was really quite big. And still the tiny creature laboured away, painting its world a brilliant crimson.

She watched, and the rim of the red dot got bigger and bigger, until she could only see a small part of it. And as it got bigger, it got straighter and straighter.

And the little ant continued painting, and continued muttering to itself, 'Painting the dot, painting the dot, red paint, red paint, round and round . . .'

Vikki was amused and contemptuous at the same time. *Silly beast*, she thought. *Doesn't it know that the world goes on for ever? It will never finish its red dot!* She didn't say this – and even if she had, she doubted the little animal would have heard, so engrossed was it in its task. And then something funny seemed to have happened, for the ant was now going round and round inside a white circle. It had painted itself *inside* a shrinking hole in its red dot.

This made astonishingly little sense.

And still the ant persevered, until finally it was completely surrounded by a vast expanse of red, and was balancing on tiptoes (not that ants have toes, but that's what it looked like) inside a white dot that was smaller than its own body. It put down its pot and brush, and its endless chant changed into 'Painted the dot, painted the dot, time for a rest, time for a rest, let the paint dry, let the paint dry; painted the dot, painted the dot, time for a rest, time for a rest . . .' Until the paint *did* dry. And then—

Then, the ant picked up its pot and brush, and began circling the white dot, chanting, 'Painting the dot, painting the dot, white paint, white paint, round and round . . .' And the white dot grew, and grew, until she could only see a small part of it. And as it got bigger, it got straighter and straighter . . .

She was intending to ask the Space Hopper what it meant, but when she woke up, she forgot. There was so much to do, so many spaces to hop . . .

The next night she had a different dream – a nightmare. She dreamed that *she* was an ant, in Spaceland this time. In one hand she held a small red ball, in another a pot of red paint, and in a third (ants have six limbs, remember) a brush. And she was singing a little song: 'Painting the ball, painting the ball, red paint, red paint, round and round, painting the ball, painting the ball, red paint, red paint, round and round . . .' And as she sang she applied layer upon layer of thick paint to the ball. The red ball got bigger and bigger until she could only see a small part of it. And as it got bigger, it got flatter and flatter.

And Vikki continued painting, and continued muttering to herself, 'Painting the ball, painting the ball, red paint, red paint, round and round . . .'

And then . . . something funny seemed to have happened. For Vikki was now going round and round inside a spherical hole. She had painted herself *inside* a shrinking hole *inside* the ball.

She was trapped!

At which point she woke up in a cold sweat.

This time, she remembered the dream, and told the Space Hopper all about it.

•

'You were dreaming you were inside a 3-sphere,' said the Space Hopper.

'You mean there's something sensible behind it? I thought my mind had just made up some kind of nonsense.'

'Not at all. Exotic dreams are a common side-effect of using a VUE. It's entirely sensible.'

'It looked just like Spaceland.'

'Yes, 3-spheres do, Vikki. If they're big enough. It's a trap a lot of people who paint balls fall into. It's amazing how often they have to be dug out with pneumatic drills.'

'What's a 3-sphere, Hopper? What was happening?'

'Let me put it this way, Vikki. The mathematics of multidimensional spaces is based on generalizations from low-dimensional spaces. For example, every point in Flatland, a 2D space, can be specified by two coordinates, and every point in 3D Spaceland can be specified by three coordinates. And a point in 4D space ought to correspond to a set of four coordinates, and a point in *n*D space ought to correspond to a list of *n* coordinates. So *n*-dimensional space itself (or *n*-space for short) can be thought of as the set of all such lists.'

'Is space really a lot of lists?'

'Mathematicians don't worry about what things *really* are. They just want to find effective ways to work out what they can do.'

'Oh.'

'Now, the same kind of mental trickery leads to formulas for distances in *n*-space, angles, and the like. From there on, it's a matter of imagination: most sensible geometric shapes in two or three dimensions have straightforward analogues in *n* dimensions, and the way to find them is to describe the familiar shapes using the algebra of coordinates and then extend that description to *n* coordinates in whatever way seems most obvious.

'A circle in the plane, or a sphere in 3-space, consists of all points that lie at a fixed distance from a chosen central point. The same idea applies to *n*-space. The set of all points that lie at a fixed distance from a chosen point is known as an (*n*–1)-dimensional *hypersphere*, or an (*n*–1)-sphere for short.'

'Why the minus one?'

'Ah. The dimension drops from *n* to *n*–1 because . . . well, because it makes sense. You see, a circle in 2-space is a curve, which is a 1D object, and a sphere in 3-space is a 2D surface. A *solid* hypersphere in *n* dimensions is called an *n*-ball. So Planiturth is a 3-ball and its surface is a 2-sphere. This kind of stuff is a bit prosaic, but once you've set it all up you can happily talk about a 9-cone in 11-space whose base is a 4-sphere and whose axis is a 5-plane. Or whatever.'

'And . . . you said I was trapped inside a 3-sphere?'

'Yes. The ant in your first dream, you see – it was really on the surface of a very big 2-sphere. It started painting at the north pole, and by the time it had worked its way down to the equator, the edge of its red dot was looking pretty close to a straight line. Actually it was a gigantic circle, but you couldn't see that, you were too close to it. Then, when the ant had painted its way south of the equator, it was painting itself inside an ever-shrinking circle, surrounding the south pole. So it waited for the paint to dry, and then it started again with white paint, working its way northwards. A very hard-working ant, was it not?

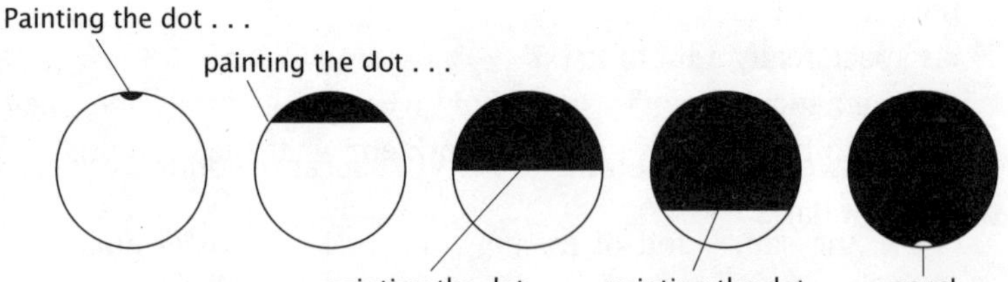

'Your second dream was what would happen if you did the same kind of thing inside a 3D analogue of a 2-sphere – that is, a 3-sphere. The ball represented its "north pole". As you painted the ball, and it got bigger, you were actually filling the northern "3-hemisphere" with paint, and eventually you passed the equator.

From then on it was only a matter of time before you got yourself trapped inside an ever-shrinking 3-ball centred on the "south pole".'

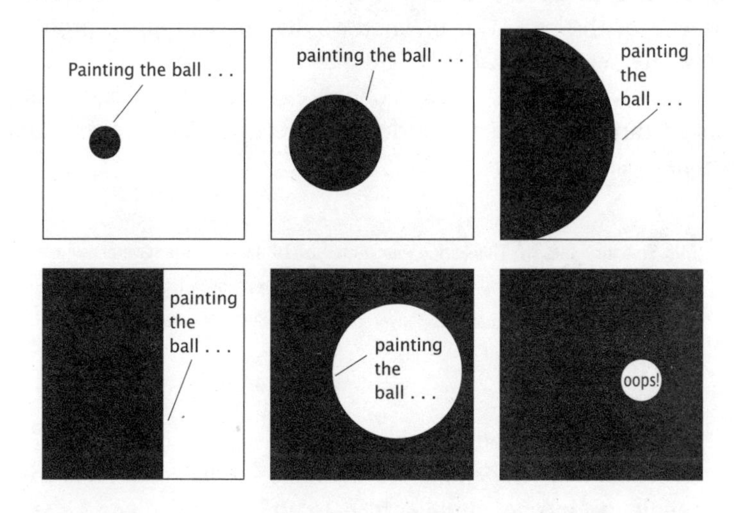

'This multidimensional geometry is weird stuff,' said Vikki. 'But it has its own kind of logic, doesn't it?'

'It certainly does. It makes so much sense that after a while you get used to the geometric language. But you can always do the algebra to *check* it makes sense, if you're worried.'

'Actually, what worries me is something quite different,' said Vikki. 'Does this multidimensional stuff let you do anything *useful*? Or is it just intellectual fun and games?'

The Space Hopper's grin became even more manic than usual. He must have done something to her VUE, because suddenly everything went dark.

•

Something went *whizzzzzzzzzz!* past her ear, and she ducked. It left a streak of blazing white light like a meteor. Moments later, another streak shot past, and another. As her eyes became accustomed to the dark, she realized that white streaks were all around her.

'Where are we?'

'We are in the Land of Dotcom,' said the Space Hopper.

'And what are those whizzy things?'

'Let's find out.' The Space Hopper produced a boxlike device and held it up. As the next white streak shot past, its trajectory suddenly bent through a hairpin curve and it sizzled straight into the end of the box. 'Got it!' shouted the Space Hopper, proudly. 'Now, let's see just *what* we've got.'

There was a little window in one side of the box, and when Vikki looked through it she saw little red letters, forming words. They scrolled past her eyes:

> Mrs Smoggrimble, HI. Thanks for your e-mail of 14 January. We have consulted our files and regret to inform you that we can find no record of any payment for the crate of poodle dye that was delivered to you. Unless the invoice is settled in full within 30 days we will HAVE TO SUE. :-)

'What's all *that* about?'

'It seems that Mrs Smoggrimble thinks she has paid for a crate of poodle dye but her supplier begs to diff—'

'I can see that. I mean: what does it signify? How does it relate to those whizzy things?'

The Space Hopper bounced delicately. 'Ah. It *is* one of those whizzy things.'

'How can a whizzy thing that streaks past at the speed of light be made of *words*?'

'Because those whizzy things are messages. Just think of them as very rapid messages, like someone who talks very fast. A used cardioid dealer, say.'

'Hopper, we've got e-mail in Flatland,' said Vikki dismissively, 'but I was asking about the uses of many dimensions. Instead of answering, you're showing me an e-mail message. Why can't you stick to the point for once?'

'I *am* sticking to the point, Vikki. Messages can be thought of as points in multidimensional space. Modern digital communications use sequence of 0's and 1's to encode information. For example, here they might encode the letter M as 10010, R as 11100, whatever. So the message you read would be coded as a string of binary digits, starting with 10010 11100. Let's just think of those first ten digits. You can think of the string as a Point in a ten-dimensional

Space, whose coordinates are either 0 or 1. The first coordinate determines the first digit, the second coordinate determines the second digit, and so on.

'Now, the trouble with digital messages is that interference can cause errors. Say the message is sent out looking like 1001011100, but arrives as 100101**0**100, where the seventh digit has changed from 1 to 0. If you get a lot of that kind of thing the message might be received as:

> Mr Snogthimble, oi. Thinks fur you re-mail of 174th canary. We have insulted your flies and piglet to infoom you that ?XX**?? grind on reword any playmate for the cart of noodle eyes that was undelivered you. Unload the thin ice is speckled in fool within 03 dogs we well. Love to Sue. :-{

'. . . which would be confusing.'

Vikki laughed. 'I can see that. So what do you do? Get rid of the interference?'

'Nope, usually impossible. But there's a clever trick: error-detecting and error-correcting codes.'

'How can you detect an error if you can't understand the message you receive?'

'If you can't understand it, Vikki, then probably something is wrong. But the idea is cleverer than that. For instance, you could detect (but not correct) any isolated error if you coded every message by replacing every 0 by 00 and every 1 by 11. Then a message such as 110100 would code as 111100110000. If this is received as 11**10**00110000, with an error in the second pair of digits, you know something's screwy because the pair 10 should never occur.'

'Oh. I see. But you can't tell whether it should have been 00 or 11.'

'I'm coming to that. Suppose you code 0 as 000 and 1 as 111. Then any isolated error can be corrected as well as detected. Suppose your received 111000101111, say. What was the message before the error got made?'

'Uh . . . 111000**1**11111. Easy! But what's that got to do with thinking of a message as a point in a multidimensional space?'

'Ah. For that, we need to journey into one tiny corner of Dotcom. I hope you don't mind feeling cramped.'

The VUE reset, the world changed. As far as she could tell, it now contained exactly four points:

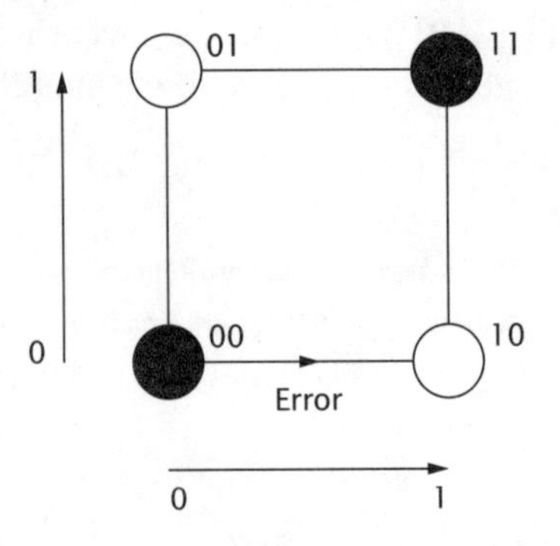

'Welcome to the Double-Digit District,' said the Space Hopper.

'This is a *district*?' asked Vikki.

'Certainly. I admit it's a bit sparse, but there's a reason. We're in a digital space. It has very few points, but, far more importantly, it has a *metric* – a concept of *distance*. Each white point is one unit away from each black point, and each point is *two* units of distance away from the other point of the same colour.'

'Don't you mean root two?' said Vikki. 'Isn't the diagonal of a square—'

'Aha! In Flatland, yes. In the Double-Digit District, no. Here diagonals don't exist – so you have to go along the edges.'

'Oh.'

'The Double-Digit District is the essence of the error-detecting code I just mentioned. Think of two-digit segments of the coded message – valid segments 00 and 11, and invalid ones 01 and 10. By thinking of the digits of these segments as coordinates relative to two axes (corresponding to the first and second digits of the segment respectively) we get the geometry of the Double-Digit District. The valid segments 00 and 11 are at opposite corners of the square, so they are *two* units of distance apart. Any isolated error changes them to segments only *one* unit of distance away, at the other two corners – but those are not valid segments.'

'Oh, I see. So because each valid segment is surrounded by invalid ones, you can tell when something's gone wrong.'

'As long as you don't get two errors in one segment, yes. But in the Double-Digit District, the invalid segments are adjacent to both of the valid ones. So different errors can lead to the same result. That's what makes it an error-*detecting* code, but not an error-*correcting* one.'

'Fine. So how do you correct errors using a code?'

'The most obvious way to get an error-correcting code is to move into Triple-Digit Territory. If we use segments of length three and encode 0 as 000 and 1 as 111, then the segments live at the corners of a "cube" in a finite 3D space. Any isolated error results in an adjacent segment; moreover, every invalid segment is adjacent to only *one* of the valid ones.'

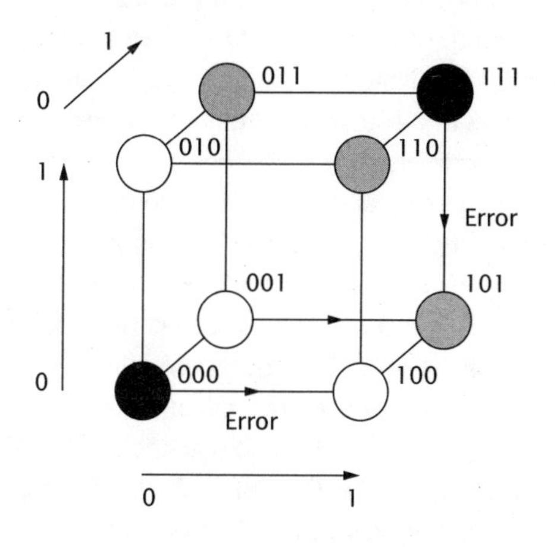

•

Fourday 5 Noctember 2099

Dear Diary,

You'd think that a Space consisting of isolated points would be exceptionally dull.

Not a bit of it. Especially the multidimensional ones. And the amazing thing is, the InterLine wouldn't work without them. Every message sent to or from a computer would be corrupted – I *like* that word, sounds very streetwise, doesn't it? – by noise.

The Space Hopper says that it all works because of a gadget called a Hamming metric. I rather assumed this was some kind of incompetent actor until he explained that it was a beautiful, simple idea for turning coding theory into geometry.

You see, on Planiturth – I've already told you why we can't really get away from that Place, have I not? – this approach to code messages was pioneered by a People named Richardhamming in 1947. In the geometric imagery that stemmed from his original algebraic ideas, codewords of length *n* correspond to the corners of an *n*-cube – a hypercube in *n*-dimensional space. Single errors correspond to moves along an edge of the *n*-cube; double errors correspond to moves along two successive edges, and so on.

The Hamming metric tells you the *distance* between two codewords – how many single-digit changes it takes to convert one into the other. In fact, it's just the distance between them measured along the shortest path *formed by edges of the hypercube*. Each edge has length 1 unit, so in the Hamming metric the opposite corners of a square are 2 units apart, not √2 units apart as they would be in Flatland . . .

Hamming hypercubes don't have diagonals, OK?

Now comes the clever part. Codewords 2 units apart in the Hamming metric can detect one error but not correct it. Codewords *3* units apart can correct one error or detect two, and so on.

Richardhamming and his successors discovered a whole *slew* of useful codes. And there's payoff in other areas of the Mathiverse, such as sphere-packing. Back to Johanneskepler and the six-cornered snowflake, OK? A lot of scientific problems, like crystal structure, boil down to finding efficient ways of packing circles in 2D or spheres in 3D. Now, you can ask the same questions in *n*D, and use the results to do things like number theory. In 1965 a Planiturthian People named Johnleech took an idea from another one called Marcelgolay, and discovered a really efficient packing of 23-spheres in 24D. No People has found a better one since! In fact, it's so efficient that it led to a solution of the kissing number problem in 24D.

No, Diary, don't be naughty. It's not like that *at all.*

The problem is this: what is the largest number of nonintersecting hyperspheres that can touch ('kiss', OK? Satisfied?) a given hypersphere? Planiturthians know the answer in exactly five cases: 1D, 2D, 3D, 8D, and an astonishing 24D. And how do they *know* such obscure factoids? Well, it's a very curious story. In 24D a whole heap of Planiturthian mathematicians managed, between them, to prove that the kissing number was at most 196,560. The proof was highly indirect and nobody knew if that number could actually occur. Johnleech's packing – amazingly – gave exactly the same number, so that was the end of the kissing number problem in 24D. (Oh, and the kissing numbers in 1D, 2D, 3D, and 8D are 2, 6, 12, and 240.)

•

The Space Hopper must have dropped off for a snooze. When he awoke, Vikki was counting.

'One hundred and ninety-six thousand, five hundred and fifty-*one* . . . One hundred and ninety-six thousand, five hundred and fifty-*two* . . . One hundred and ninety-six thousand, five hundred and fifty-*three* . . . One hundred and ninety-six thousand, five hundred and fifty-*four* . . .'

'Vikki, what are you—'

'*Don't interrupt!* One hundred and ninety-six thousand, five hundred and . . . and . . . Um. Er, *one, two* . . . Bother!'

'You don't need to check it so laboriously,' the Space Hopper pointed out. 'The Planiturthians have calculated it, I'm sure they got it right.'

'I just wanted some practice with the VUE. You know, this gadget almost makes the Mathiverse *too* easy.'

'Except when you lose count.'

'Shh! That was a *joke*. I saw you were asleep and I started from one hundred and ninety-six thousand, five hundred and forty. But what I'm trying to say is, how would anyone visualize multidimensional spaces *without* a VUE? How do Planiturthians manage, for instance? Do they have VUEs?'

'Not yet. They're working on them. They're up to Virtual *Reality*. Virtual Unreality is some way off, as yet.'

'But this Johnleech People thingy must have . . . I mean, he surely couldn't have—'

'Ah. No, he didn't get the kind of image you perceive through the effortless medium of the VUE. *He* had to *think*. He had somehow to equip himself with *n*D spectacles. I don't know how he did it – probably a lot of algebra but thought of as geometry. But Planiturthians have taught me a few simple tricks, little more than analogies with 2D and 3D, which go a long way.

'Let's . . . I know, think of your ancestor! No VUEs in *his* day, I assure you. Yes, suppose that great-whatever-grandad Albert Square is sitting happily in Flatland, and wants to "visualize" a solid sphere. How does he do it?'

'Well, in his story he talked about the Sphere passing through the plane of Flatland, and moving perpendicular to it, so that what he saw was a series of cross-sections of the sphere.'

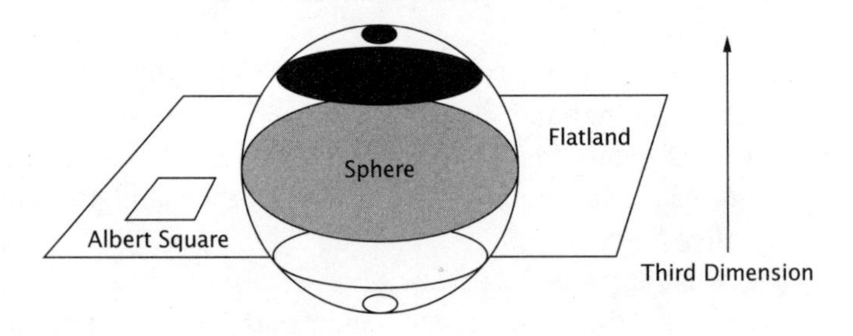

'Good! Yes, first he sees a point, which grows to form a Circle. The Circle expands until he is seeing the equatorial section of the sphere, after which it shrinks again to a point and then vanishes.'

'Well – look, actually old Albert saw the Circles edge-on – line segments with graded shading – but his visual senses *interpreted* this image as a Circle.'

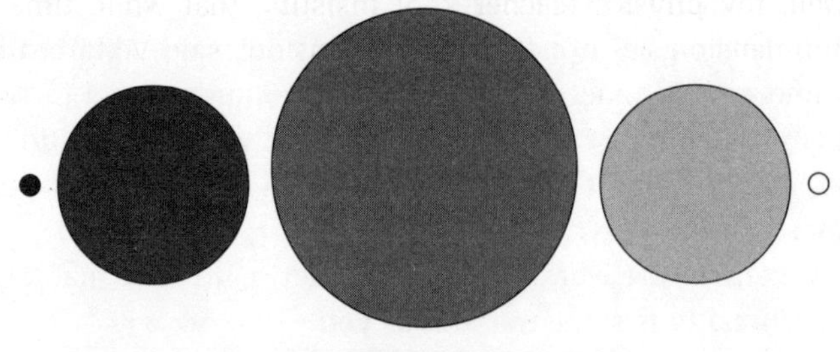

Albert sees a 'movie' of changing discs

He can also think of this as a static sequence of colour-coded frames

'Don't be pedantic. Anyway, by analogy a Planiturthian can "see" a 4D-ball – a solid hypersphere in 4D-space – as a point which grows to form a solid 3D-ball, expands until it reaches the equator of the 4D-ball, then shrinks back to a point, and finally disappears.'

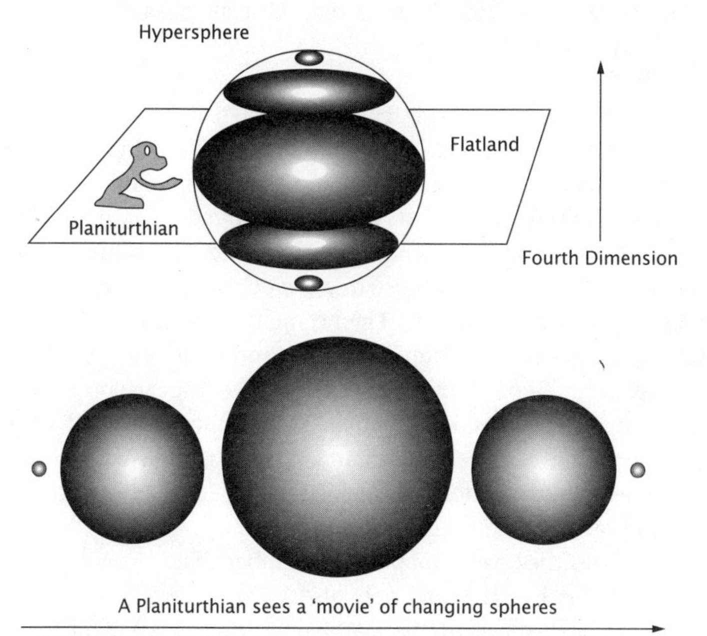

A Planiturthian sees a 'movie' of changing spheres

He can also think of this as a static sequence of colour-coded frames

'So in both cases the extra dimension is really represented as *time*?'

'You could say that, Vikki, yes.'

'Well, my physics teacher kept insisting that while time is *a* third dimension, it's not *the* third dimension', said Vikki heatedly. 'Mind you, he also said that there was no *real* third dimension at all, and thanks to the VUE I now realize that's nonsense. Nevertheless—'

'Nevertheless, using time is all very well, but the problem with time is that it has a life of its own. I met Time once, and it definitely did have its own life, I can assure you. Quite an interesting one, but *it could never go back.* Time is always moving in one direction – into the future. What we want is what Herbertgeorgewells's Time

Traveller allegedly invented – a way to move at will up and down along some substitute for a fourth dimension.

'One of the Planiturthian tricks is to imagine a fourth dimension as a *colour*. Imagine displacement along a fourth dimension to be colour-coded, say from blue to red, with all possible shades of purple in between.'

'Why?'

'The advantage of using colour is that it doesn't imply some particular physical quantity, such as time. Planiturthians are very used to the idea that *any* quantity can be colour-coded.'

Vikki could relate to that. 'Same on Flatland. Well, depends on the era. Colour was freely available at the time of Chromatistes and the Colour Revolt, and in those days colour was used with tremendous freedom and creativity. Then came the backlash, of course. It was the stupidity of the Revolutionists and their Universal Colour Bill that did it. Typical totalitarian attempt to impose some kind of uniform code of practice – and of course they got it all wrong. Imagine trying to have Women and Priests painted with virtually the same colours! Endless confusion, and it led to a massacre with the Suppression of the Chromatic Sedition.'

'Yes,' said the Space Hopper. 'I remember that.'

'*You* remember! How old *are* you?'

'In the metaspace of the Mathiverse, time has no meaning. I have no age – I simply *am*.'

'Oh. Anyway, the idea of using colour as a form of decoration couldn't be suppressed for ever, and much later, after the Six-Year War, some daring young women pioneered a new colour-conscious fashion. We've never really looked back, actually. Anyway, what I'm really saying is – you're right. Colour is a pretty arbitrary way to represent something.'

'I'm glad you agree, Vikki,' said the Space Hopper. 'Now, where was I – oh, yes. Anyone wishing to visualize a fourth dimension, say, can mentally equip themselves with a Colour Machine instead of a Time Machine. Merely by depressing a simple pedal they can imagine themselves moving at will along the red–blue axis.

'At each stage they "see" a 3D cross-section of 4-space corresponding to that colour. As the colour varies, they can mentally "stack" these images. Different sections that seem to overlap in 3D but have different colours don't really overlap, because they exist in different "levels" of 4-space. In the same way, two Planiturthian cars that pass through a crossroad will collide only if they occupy the same position *at the same time.*'

'So . . .' said Vikki, 'if Albert used a Flatland Colour Machine to think about a 3-ball, he would imagine a blue point which grows to a disc and gradually becomes tinged with more and more red. The disc reaches its maximum size when it is a beautiful purple, 50 per cent blue and 50 per cent red. As it reddens further the disc shrinks, and eventually becomes a red point, after which it vanishes.'

'Correct.'

Vikki had a moment of inspiration. 'Hopper! I've just realized! A Planiturthian could visualize a 4-ball in the same way! They start at the blue end, where they see a single point. As they depress the pedal, moving in the red direction, that point turns into a purplish-blue ball, and gets bigger. When they reach the halfway mark they see a solid purple ball. Further into the red, the ball shrinks again, and turns redder. Finally they see a single red point, which then vanishes.'

'Very good. Now, it is well known that Planiturthians have problems visualizing knots. Can you tell me how they could use a Colour Machine to visualize how to untie a knot in 4D *without breaking the loop*?'

'Um. I guess they . . . yes, yes, I can see! When Planiturthians look at a knot, they see places where the string seems to cross itself. It doesn't really, of course – that's just an effect of perspective. But those apparent crossings tell them the shape of the knot. At each crossing, one bit of string goes underneath – I'll call that bit an "underpass". And another bit goes over the top – an "overpass". Anyway, what the Planiturthians ought to do is imagine a blue loop, tied in a trefoil knot, say. Near one of the crossings they imagine grasping the overpass bit of the loop, and they pull that bit out *along the red direction.*'

The Space Hopper bounced gently as he contemplated this idea. 'Yes. So nearby sections of the loop have to turn various shades of purple, in a graduated way, to maintain continuity?'

'Sure, Hopper, sure. But most of the loop can stay blue. Now they push the purple overpass down *through* where the underpass is. It *looks* like they intersect, but actually they don't, because the underpass *is still blue*! Things that intersect must do so at a point with a common colour (the fourth coordinate) as well as a common spatial position (the other three coordinates). And now they can safely release the colour-pedal and return that loop back to the blue region of 4-space. The end result is as if they'd just pushed the overpass down through the underpass. So the loop is no longer knotted.'

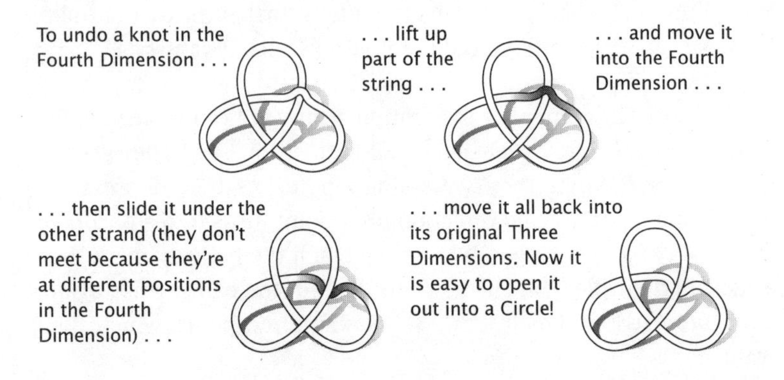

•

Fiveday 6 Noctember 2099

Well, Diary Mine, multidimensional Spaces are a lot of fun, and coding theory shows they can be useful.

And it doesn't stop there.

According to the Space Hopper, a solar system or an economy can be viewed as a multidimensional space. I admit it sounds unlikely: the Planiturthian solar system, for instance, is just a lot of rather big rocks trundling around in ordinary 3D space. But the Space Hopper says that appearances can be deceptive – or rather that they can be absolutely correct, but irrelevant.

The way he sees it, every body in the Planiturthian solar system needs three coordinates to specify its position. So with nine planets, that's 27 coordinates – and anything that has 27 coordinates necessarily lives in a 27-space! In fact you really need a 54-space, because positions alone don't tell you what the planets are doing. You need to know their *velocities*, and that's another 3 coordinates per planet.

And if you include the asteroids . . . well, you're looking at TENS OF THOUSANDS of dimensions!!! And, that, he says, is what makes celestial mechanics so hard.

I thought he was joking, but he's deadly serious. 'The Curse of Dimensionality,' he calls it.

So, Big Hairy Deal, a solar system has oodles of dimensions. Kind of esoteric, though. But – get this – an economy has *millions* of dimensions! Each dimension corresponds to the price of some commodity – say the first dimension is the price of oxagons, the second the price of hogsagons, the third the price of squarrel-feed, and so on. And that, he tells me, is what makes economics even harder than celestial mechanics.

No, that's not meant to be a joke, either.

Ah, Diary my Old Friend, I know what you're thinking. MultiD geometry provides an elegant reformulation, yes – but does it offer any practical advantages?

The Space Hopper says 'yes'. He says that, thinking about economic problems multidimensionally led a Planiturthian called Georgedantzig to some weird idea called the 'simplex method', which can find the most profitable combination of goods that a factory ought to produce. Allowing possessors of said method to make oodles and oodles of money!

Could physical space *really* have more than three dimensions? After all, how can you fit the fourth dimension in? Everything's filled up already.

Well, of course that's what Ancestor Albert's book was about. Physical reality has many interpretations, and the one presented to us by our sense organs need not be the only possibility. Our universe could be just one of an infinite variety of universes that might have happened instead, so that we occupy just a wafer-thin slice of a vast multiverse.

Mindboggling, isn't it?

•

'So now we've visited all the possible spaces?' asked Vikki.

'Whatever makes you think that?'

'Well, Hopper . . . we've done 0D and 1D and 2D and 3D and 4D and all the way up to a millionD, and I can kind of see how it would go from there and I don't think we really need to visit the rest. What else is there?'

'There's a lot more to spaces than just their dimensions,' replied the Space Hopper. 'But even if we stick to dimensional variations, there's plenty more to find out about. What about one-and-a-quarter-dimensional space, for a start?'

Vikki stared at him as if he were mad. 'But you can't possibly have one and a—

ONE AND A QUARTER DIMENSIONS

'—quarter dimensions!' said Vikki in bewilderment. But even as she spoke, she experienced that by now so familiar psychic discontinuity, and her VUEpoint shifted. There were times when she wished the Space Hopper wouldn't take her every statement as a challenge, to be demolished by conveying her to yet another weird space. However, it seemed impossible to stop the fat bouncy little creature with the thick orange skin from reacting to questions with actions. It was certainly a striking teaching technique, high on drama and nervous energy – but there were times when she would have preferred a more relaxed approach.

She stared through the nonexistent realm of metaspace as if it weren't there (which it wasn't, not in any sense of *there* that could be understood by mortal creatures). Where metaspace wasn't, there seemed to be – a forest.

'The Fractal Forest,' the Space Hopper announced proudly.

'It's very . . . beautiful,' said Vikki. And it was. There were gnarled trees, and spiky-leaved bushes, and thick clumps of fern. Snow was falling, and every surface was picked out in soft, diamond-studded icing.

'So what's with the forest, Hopper?'

'Look at the fern, Vikki.'

She tugged at a frond. 'It's . . . Hopper, it's just an ordinary fern. Intricate, pretty – so what?'

'Use the VUE. Look *closely*.'

'Well, what I see is a lot of fronds lined up on both sides of the stalk . . . and each frond is *also* a lot of smaller fronds lined up on both sides of a tinier stalk. That's what makes ferns ferny, isn't it?'

'Closer than that.'

'OK . . . each smaller frond is a lot of even smaller fronds lined up on both sides of an even tinier stalk . . . and . . . oh. This one doesn't stop, does it?'

'No. It goes on doing that for ever. It's not a *real* fern: it's a mathematical idealization of fern structure. The trees and bushes have the same infinitely fine repeating structure. And so, oddly enough, does the snow.'

Vikki readjusted her VUEfinder. Hovering in mid-air, sparkling and crisp in the unremitting light of VUE-enhanced imagination, was a single snowflake.

It was the sort of snowflake that would appeal to a Flatlander, made – so far as she could tell – from an endless series of ever-smaller equilateral triangles. There was a big triangle, with three smaller ones placed on its sides to make a six-pointed star . . . then there were (she counted) twelve smaller triangles, placed on the edges of the star to make a more elaborate star. *Then* there were (she lost count five times and gave up) a *lot* of even smaller triangles placed on the edges of that new star. When she Cranked up the magnification of her VUE it became clear that the sequence of triangles didn't stop there: the closer she looked, the more of them she saw. The edge of the snowflake was crinkled, and there were crinkles on the crinkles, and crinkles on the crinkles on the crinkles.

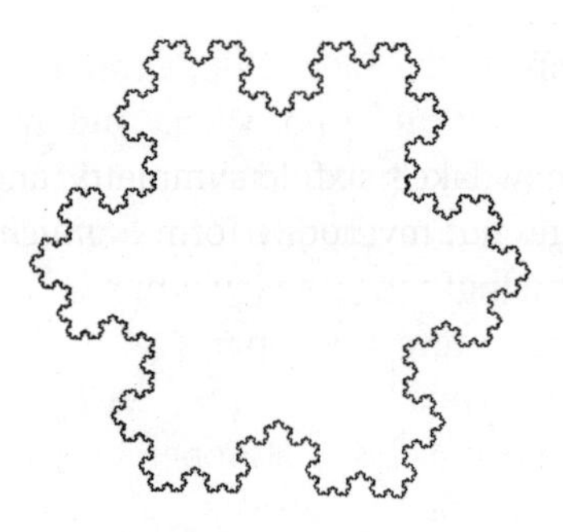

The snowflake spun gently in the brilliant light, like a tossed pancake in low gravity. 'Enough of that', said the Space Hopper, and twiddled some extradimensional knob or lever or metaspatial demon, freezing the snowflake in one position. 'In the Fractal Forest, geometry goes on for ever and dimensions come in fractional amounts. The edge of this snowflake, for instance, has *approximately* 1.26186 dimensions. It's not *exactly* one and a quarter, I accept, but if you really want I can certainly show you a shape that—'

'Whoa there, Hopper! Let's just back up, OK?'

'Sorry? You have a problem?'

Vikki tried to pull her scattered wits together. She had a feeling that only part of her discomfort was the psychic dislocation caused by the Virtual Unreality Engine. 'Look, dimensions can't be fractions.'

'Tell that to Helge the Snowflake, then,' said the Space Hopper. A quizzical eye opened at the centre of the six-pointed star. 'Sorry, Helge, but you've got another unbeliever.'

'Just my luck,' said the snowflake glumly, 'I should have known. It always happens to me. Every Tim, Dick and Harriet who wanders into the Fractal Forest always ends up on my doorstep sooner or later – and it's usually sooner.'

'It's because of your inherent simplicity and elegance,' said the Space Hopper, 'plus your parents chose an attractive name.'

The snowflake's edges fluttered in acknowledgement of the compliment. 'I suppose so. But it's still a great burden to have to bear, you know. If I had a penny for every time my circumference has been measured, I'd be able to retire to some nice fractal forgery of an island and spend my time sunbathing on its infinite beaches.'

'Are you a *real* snowflake?' asked Vikki.

'No, no, not at all. If I were, then I wouldn't be able to sunbathe – I'd melt.' The snowflake paused, as if gathering its wits. 'However, I do share a real snowflake's sixfold symmetry, and if anything I am even more intricate, but my bodily form is *much* too regular for a true snowflake. Think of me as a visual pun.'

'The Space Hopper said you had 1.26186 dimensions,' said Vikki. 'I think that's rubbish.'

The snowflake scratched its edge with one frilly corner. 'I think someone discovered that rubbish has about 2.8739 dimensions,' it told her, apparently in all seriousness. 'Of course, it does depend on how you crush it, but—'

'I meant, you were *talking* rubbish.'

The snowflake looked offended (though you'd have to be an

expert to spot such nuances of expression). 'Me? Rubbish? Perish the thought! You'll get nothing from me but the most sparkling insight. But *my* dimensions are two, just like many of your fellow citizens. You, of course, are one-dimensional.'

'But—'

'Ah! The dimensions of my *boundary* are 1.26186, whereas your boundary, if I'm not mistaken, has exactly zero dimensions. Let me check . . .' (it began counting on its vertices) '. . . one, two, three . . . mmmm . . . mmnnnhmmm . . . hmmmhmmm . . . *zero*. Right.'

'My boundary consists of two points', said Vikki.

'Correct! And a point is zero-dimensional, and the same goes for any finite set of points. Including two.'

'Zero dimensions, I can understand. It's really just a convention – it fits the sequence cube/square/line/point. But 1.26186 dimensions? That's nonsense. The number of dimensions something has is the number of different *directions* it points in – the number of numbers needed to specify a point in it. You can have one direction, or two, but you can't have 1.26186 directions. Can you?'

'Of course not, Vikki,' said the Space Hopper. 'But you know what mathematicians are like – always meddling. No sooner does one of them come up with a definition of dimension in terms of directions, when some other smartarse has to improve on the idea by finding a completely different definition that gives the same answer when the dimension of a space is a whole number, but works for other spaces too. More general, you see, which to a mathematician means "better". Except that with the *new* definition, you can get fractions.' It swelled slightly with excitement. 'They were Planiturthians, those mathematicians . . . Can't say I was surprised. Whole-number dimensions . . . well, to be truthful, those had been around since the time of Euclidthegreek, at least in principle. But a Planiturthian called Henripoincaré pushed the idea about as far as it can be pushed. According to Henripoincaré, *every* shape, however weird, has some whole number of dimensions. As long as you allow infinity, of course . . . oh, and minus one.'

'*Minus one?*'

'Sorry, don't be nervous. That's just a convention. The only –1D set is the empty set.'

'Which is?'

'Nothing at all. If you draw a picture of it, it's blank.'

'Let me see if I'm getting this right,' said Vikki. 'According to this Hairypunkarray—'

'Henripoincaré.'

'—every shape has a whole number of dimensions.'

'Correct.'

'But you just said that Helge the Snowflake has 1.26186 dimensions.'

'No, that's my *boundary*—' Helge protested.

'Boundary, schmoundary. Whatever, it's not a whole number.'

The Space Hopper bobbed in agreement. 'Yes, but that's because another Planiturthian disagreed with Henripoincaré. *His* name was Hausdorffbesicovich, and *he* had quite a different idea about dimensions. He reckoned that it wasn't a matter of more dimensions meaning more directions – it was more dimensions meaning filling space more effectively. More crinkly, so to speak. You might call it "crinkliness" instead of "dimension", but of course mathematicians wouldn't do that, because the name wouldn't sound impressive enough. And, to be fair, there's another reason. You see, on all the usual spaces like the 2D plane or 3D space or even 101D space, the Henripoincaré-dimension and the Hausdorffbesicovich-dimension are the *same* – they're 2 or 3 or 101, as you'd expect. So both notions *generalize* the "number-of-directions" version of dimension.'

'Unfortunately,' said the snowflake, 'they generalize it in two incompatible ways.'

'Yes – but that's maths for you. As far as mathematicians are concerned, the more kinds of dimension you invent, the happier they are. And if some kinds are incompatible with one another, you can have a lot of fun finding out why and when. So now there's the box-counting dimension, and the similarity dimension, and the information dimension, and—'

'*Stop!*' Vikki lowered her voice to more normal levels. 'You've already told me enough dimensions to last me a lifetime. But you still haven't explained how Helge can have 1.26186 dimensions.'

'Haven't I? Dear me, no, I haven't.' The Space Hopper was momentarily embarrassed. 'Uh . . . well, the easiest way to see *that* is to use the similarity dimension.' The Space Hopper waved his horns like a conjuror, and a piece of card materialized between them. He laid the card flat (in metaspace, you appreciate) so that it seemed to be hovering in thin air. With a flourish, an entire pack of similar cards appeared, and the Space Hopper fanned them out. 'Now, Vikki: how many of these do I need to make the original square twice as big?'

'In what sense?' she queried, having learned caution. The Space Hopper asked too many trick questions!

'By fitting them together. Look, let me show you the answer and you'll see what I mean.'

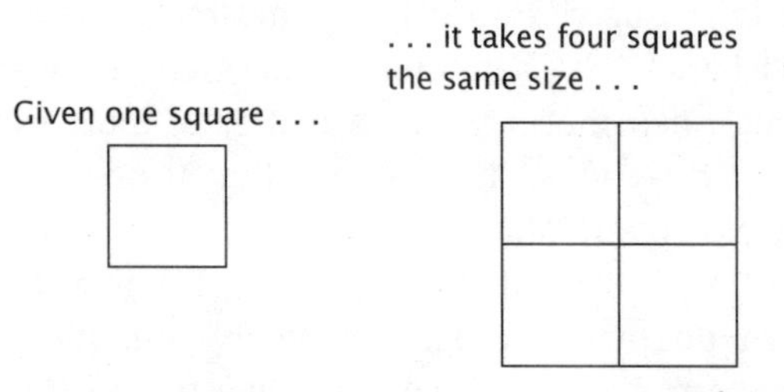

'Right, I get it,' said Vikki.

'It works with triangles, too. Four copies *also* make a triangle twice the size.'

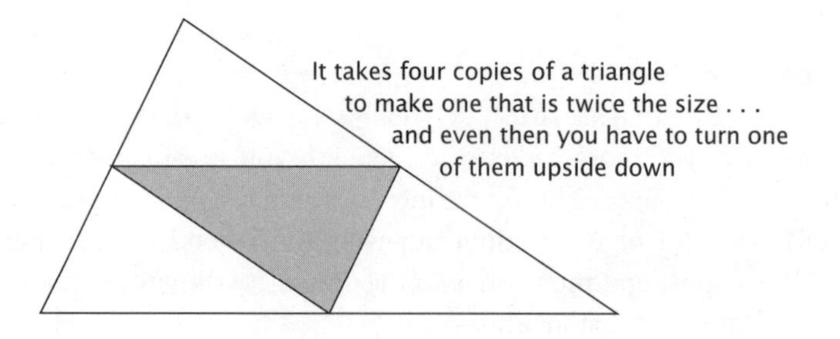

'Of course.'

'What about cubes?'

'Cubes?'

'Do four cubes make a cube twice the size?' The Space Hopper twitched his horns, and four identical cubes materialized in her VUEfield.

'Um . . . no. They do make a sort of square slab, but it's too thin to be a cube.'

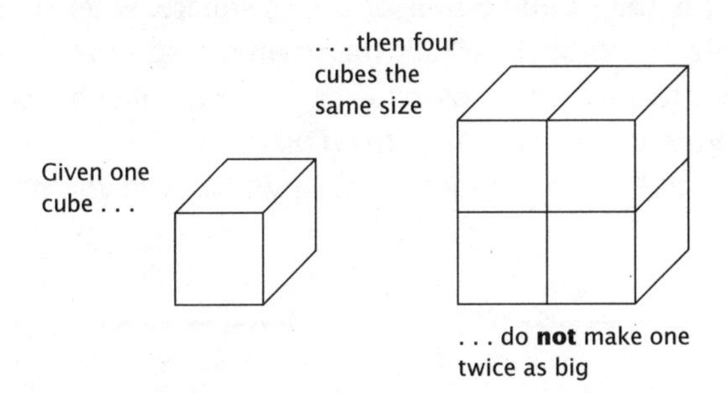

'So how many more—'

'Four! If we make a *second* slab, and fit it behind the first one, we get a cube twice the size of the original.'

'Excellent. But why?'

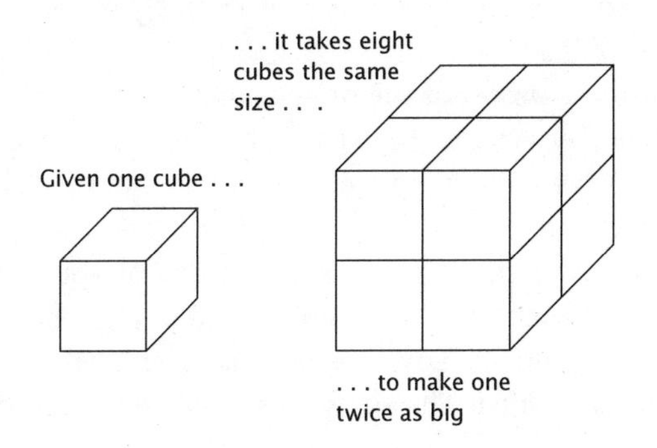

'Well, a cube's thicker than a square, so you have to have two layers to double the thickness. A square has no thickness, and twice zero is zero, so one "layer" is enough.'

The Space Hopper grimaced. 'Yeeesss . . . but there's a better way to say that. What do you mean by "zero thickness"?'

'Um . . . it's not thick.'

'True. But in what *direction* isn't it thick?'

'The one in which it's . . . *oh*, I see what you're getting at. The cube has an extra *dimension*, which I've been calling "thickness", but the square doesn't.'

'Precisely. The cube is 3D, but the square is 2D – like the triangle.'

'I see.'

'Now, 4 is the *second* power of 2 – its square. Whereas 8 is the *third* power – its cube. So what's the common pattern?'

'In order to double the size of a shape, the number of copies you need is 2 *raised to the power of the dimension.*'

The Space Hopper grinned manically (∪). 'So if the dimension is 1?'

A line can be made twice the size by joining two copies together

'You need 2 to the power 1, which is . . . 2. *Two* copies. Oh, right!'

'And if we wanted to multiply the size by three, how many copies would we need? Here's a hint: three copies for a line, nine copies for a square, and twenty-seven copies for a cube. So the general pattern this time is?'

'3 raised to the power of the dimension.'

'Fine. Which, by a roundabout route, brings us to the snowflake.'

'I'm glad you haven't completely forgotten about me', Helge the snowflake muttered.

'Not at all, my friend. Show Vikki here one of your edges. Not your entire perimeter, that's a bit complicated. Just one section of it.' The snowflake did so. 'Lovely and crinkly, isn't it?'

'Beautiful,' said Vikki. The snowflake blushed with pride.

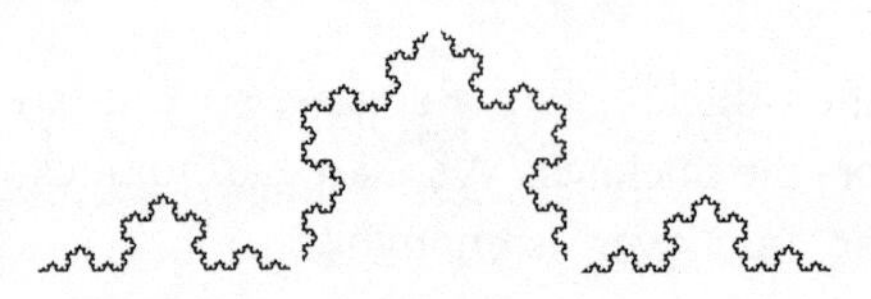

'And the interesting thing is, you can make that crinkly edge three times the size by fitting *four* copies together. Yes?'

'Absolutely. So?'

'What power of 3 makes 4, Vikki?'

'Well, um . . .' She racked her brains, and then realized it was another of the Space Hopper's trick questions. 'There isn't one.

The first power of 3 is 3, which is too small, and the second power of 3 is 9, which is too big. And after that, they just get bigger still.'

'What about the one-and-a-halfth power?'

'That doesn't – oops, yes, it does. We did this in maths. The halfth power is the square root. So the one-and-a-halfth power is . . . the cube of the square root?'

'Correct. And that is?'

'Sorry, I'm not that good at mental arithmetic.'

'OK, I'll tell you: it's 5.19615. Roughly. But unfortunately, that's—'

'Still too big. We want to get 4.'

'OK. So what about the one-and-a-quarterth power?'

'Um . . . that's the fourth root of the fifth power, right?'

'Right. Value 3.94822.'

'Which is close.'

'As you say. In fact, we can get a lot closer. The 1.26186th power of 3 – which you could work out as the 100,000th root of the 126,186th power of 3 – is *extremely* close to 4. And by going to more decimal places, you can get as close as you like. The exact number, in fact, is log 3 divided by log 4, where "log" is the logarithm.'

'So you're saying that—'

'The dimension of Helge's edge is very close indeed to 1.26186. Like I told you.'

There was a long silence. 'It's not *really* a dimension, is it?' asked Vikki. 'It's just come sort of measure of – crinkliness – that happens to be equal to the usual dimension when you have a simple sort of space.'

'Fair enough,' said the Space Hopper. 'Complicated shapes like the snowflake or the fern, whose "dimension" in the crinkliness sense is a fraction – or to be strict, just differs from its dimension in the Henripoincaréan sense – are known as *fractals.* Roughly speaking, a fractal is a shape that has detailed structure, no matter how much you magnify it. Physical examples are rocks, clouds, trees, coastlines – as long as you don't magnify them to the scale of atoms, of course. Mathematical fractals can be scaled up for ever and still have intricate structure; real fractals can be scaled up *a lot* and still have intricate structure.

'And if you want to be very precise, you can always call the crinkliness measure the *fractal dimension*. Then there's no danger of getting confused and trying to find one-quarter of a direction.'

•

'Fractals must be very rare,' said Vikki.

'Why?' asked the Space Hopper.

'All that detail. Unusual.'

'Funnily enough,' the Space Hopper told her, 'it's pretty much the other way round! Nearly all mathematical shapes are fractal. Having a whole number of dimensions is what's rare. Think of it this way: if you pick a number at random – a decimal, not necessarily a whole number – then the chance of it *being* a whole number is very small indeed. Most numbers have something after the decimal point.'

'True – I hadn't thought about it that way. So what do some of the other fractals look like?'

'I'm sure the Snowflake would be happy to let us meet some of his friends here in the Fractal Forest.'

'Delighted to be of assistance,' said the Snowflake. 'You can usually find them hanging around among the leaves and in between the tree-roots . . . Yes, here's a very close friend of mine. Like me, he's based on triangles.' And he introduced them to Sierpiński Gasket, who was made out of triangles with triangular holes, which cut it into more triangles, also with triangular holes. He soon dug up Sierpiński Carpet, who was very similar but with square holes instead. And an energetic search revealed the very impressive Menger Sponge, who was like the Carpet but made from a cube.

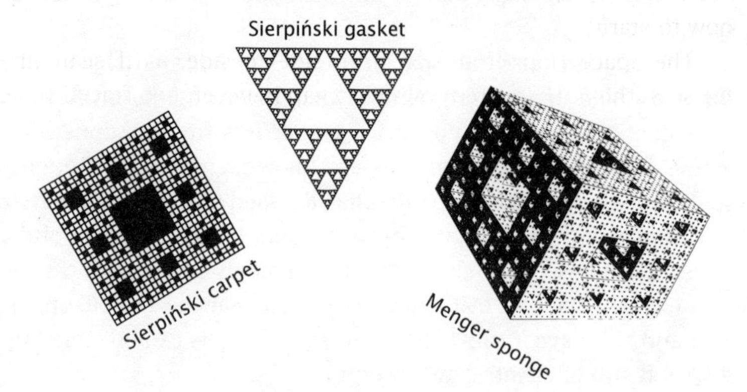

'Fractals like these,' the Space Hopper explained, 'are *self-similar*. Like a fern, they're made up from lots of smaller copies of themselves – that's what makes it easy to work out their fractal dimension. Have a go!'

'Um . . . Sierpiński Gasket is made from three copies, each half the size?'

'That's right. So his fractal dimension is log 3/log 2 = 1.5850. Whereas Sierpiński Carpet is made from eight copies, each one-third the size—'

'—so his fractal dimension is log 8/log 3 = 1.8928. I can do it too! Good job my Personal Disorganizer has a built-in scientific calculator!'

'And Menger Sponge is made from 20 copies each one-third the size, so his dimension is log 20/log 3 = 2.7268.'

'I see what you mean: the bigger the fractal dimension, the "thicker" the shapes seem to get.'

'That's the general idea, and it's what gives intuitive meaning to the fractal dimension.'

Vikki watched as the Snowflake, Gasket, Carpet, and Sponge wandered off and disappeared behind the infinitely crinkled vegetation of the Fractal Forest. 'Are all fractals self-similar?'

'Definitely not,' said the Space Hopper. 'Those are just the easiest ones to understand. But there are lots more. They have a fractal dimension too, which is usually not a whole number, but it's nothing like as easy to calculate.'

'It must be very difficult to find a shape that has structure on *all* scales,' said Vikki, 'especially when it's not self-similar. I don't see how to start!'

The Space Hopper turned from side to side, as if searching for something. He peered towards a gap between two fractal pine-trees.

'What are you looking for?'

'I'm trying to remember the way to Quadratic City.'

'Why?'

'Because there's somebody there who will show you how to specify a shape that has structure on *all* scales,' said the Space Hopper. 'OK, I think I've got my bearings. Follow me!' and he bounced off erratically through the trees.

'Who?' asked Vikki, wriggling along behind him.
'The Mandelblot.'

•

Sevenday 8 Noctember 2099
As we made our way through the beautiful landscape of the Fractal Forest, Diary Mine, the Space Hopper told me the story of the Mandelblot.

For the record, here it is.

Seems that the Mandelblot was originally called the Gingerbread Boy. Baked from gingerbread, but unusually articulate for a cookie. He changed his name when he grew older. After he was first baked, he ran away. He came to a river, and there was a fox snoozing beside it.

'Please, Mr Fox, take me across the river,' said the Gingerbread Boy.

'I am the Fox Pup, and the wise person listens to me,' said the Fox, who had an eye for a bargain, especially if the bargain was a free lunch. 'Hop onto my back and I'll swim across.'

So the Gingerbread Boy hopped onto the Fox Pup's back. But as the fox swam further out into the river, he began to sink. 'Hop onto my head,' said the fox, 'and you'll keep dry.' So the Gingerbread Boy hopped onto the fox's head. But still he sank. 'Hop onto my nose,' said the fox, 'and you'll keep dry.' So the Gingerbread Boy hopped onto the fox's nose.

And as the fox sank further, the Gingerbread Boy hopped onto the very tip of the fox's nose. And then the tip of the tip, and the tip of the tip of the tip . . . until the fox began to get very frustrated. 'How long do you intend to keep this up?' the fox demanded, in an irritated voice. 'How do you expect me to get lunch if you keep hopping onto ever finer scales of my nose? I may have to dip my nose completely into the water if this carries on.'

But the Gingerbread Boy kept hopping on to the tip of the tip of the tip of the tip of the . . . tip of the tip of the tip of the fox's nose. Because he knew that the fox wouldn't submerge his whole nose in the water. *Why not?* you are asking, Diary Dear. Because he didn't like soggy gingerbread. And the fox got seriously frustrated, and again he asked the Gingerbread Boy how long he intended to keep hopping along his nose, and when he planned to get to the end (where, of course, the fox could open his mouth and gobble him up).

'I'm awfully sorry, Mr Fox,' said the Gingerbread Boy, 'I'm afraid I'm never going to reach the end of your nose. You see, before I can get to the end, I have to get halfway to the end. And before I can get halfway to the end, I have to get a quarter of the way to the end. And before I can get a quarter of the way to the end, I have to get an eighth of the way to the end. And so on. So I can't even get started—'

But the Fox Pup had heard that one. 'Listen to me, young man,' he told the Gingerbread Boy sternly, 'I've hung around the Mathiverse a lot, and I've come across Zeno's paradox before. I don't believe a word of it. For a start, you're *already* nine-tenths of the way along my nose, and if your argument were correct, you wouldn't have been able to get that far.'

'Oh, all right, then,' said the Gingerbread Boy, 'let me put it this way. After I've got nine-tenths of the way along your nose, but before I can get to the tip, I have to get ninety-nine hundredths of the way along your nose, yes?'

'Well . . . yes,' admitted the Fox Pup.

'And after I've got ninety-nine hundredths of the way along your nose, but before I can get to the tip, I have to get nine-hundred-and-ninety-nine thousandths of the way along your nose, yes?'

'Well . . . yes,' conceded the Fox Pup, who hadn't encountered this variation before.

'So it's clear I can *never* get to the tip of your nose.'

'Let me get back to you on that,' said the Fox Pup. By then he had nearly reached the far shore, though he was too busy counting on his paws to notice, and the Gingerbread Boy trotted along to the very tip of the fox's nose and hopped off.

'Hey!' complained the fox. 'I thought you said you couldn't do that?'

'I lied,' said the Gingerbread Boy. 'The sequence is convergent.'

•

Vikki waited, but the Space Hopper had finished.

'Was that *it*?'

'Yes.'

'That's a very stupid story!'

'Well, it was a very stupid fox. Anyway, I was just passing time until I could locate the – and there he is! Hey, Mandelblot!' And the Space Hopper bounced up and down in excitement.

The Mandelblot was a very funny shape, a cross between a cat lying on its side and a cactus:

The Mandelblot . . .

. . . and a close-up

And the closer Vikki looked, the more complicated the Mandelblot's shape seemed to be.

'He's a fractal?'

'Indubitably. One of the best. But no, to anticipate your next question – he's not self-similar.'

'Why is he that shape?'

'He was made Taxi Controller, in charge of Quadratic City's taxi service, and he tells the taxi drivers whether they can get out or not.'

'That,' said Vikki, 'is as clear as mud.'

'It's quite a long story,' said the Space Hopper, 'but better than the one about the Fox Pup and the Gingerbread Boy, I promise. Come with me, and I'll ask the Mandelblot to explain—'

'Everything in Quadratic City,' began the Mandelblot, 'is based on squares. 'Why, the city itself is nothing more than a square grid.'

'How big?' asked Vikki.

'Well, from the *outside* it's about the size you'd expect for a city, but on the *inside* it's infinitely big. It's a plane, just like your home-world – yes, I can see you're a Flatty, we don't often get nonfractal folk like you in the Fractal Forest, you know.'

'Does it have pentagons and circles and things like that?' It wouldn't be the *same* as going home, but it might substitute. She still missed her family and friends – she really must persuade the Space Hopper to overcome those alleged system incompatibilities and find a way for her to phone home, or at least send a letter . . .

'Not as such, no. Mostly it has roads . . . and taxis. And taxi-drivers, of course. Both of those are points.'

'Oh.' Not at all like home, then.

'The roads are always straight lines, and there are two kinds. Streets run east–west, and Avenues run north–south. There's exactly one Street and one Avenue for every point in Quadratic City: they form a coordinate grid. So, for instance, if you get a taxi on the corner of 2.4 Avenue and 1.7 Street, and want to go to the corner of 5.6 Avenue and 3.8 Street , then the taxi-driver immediately knows that he has to go 3.2 kilometres east and 2.1 kilometres north.'

'That's a neat idea,' said Vikki. 'Fix it so's the street names tell you where you are.'

'Yes. Mind you, most of the names of the Streets and Avenues are pretty long. I *hate* it when I have to go to π Street or $\sqrt{2}$ Avenue. But that's a minor inconvenience when you consider how easy it is to navigate.

'The two main thoroughfares are 0 Street, otherwise known as Real Road, and 0 Avenue, the Queens i Way. That's "i" as in the Square Route of Minus One, you appreciate. They cross at Grand Central Station, the exact centre of Quadratic City. Taxis start out from Grand Central Station, and try to get out of the city.'

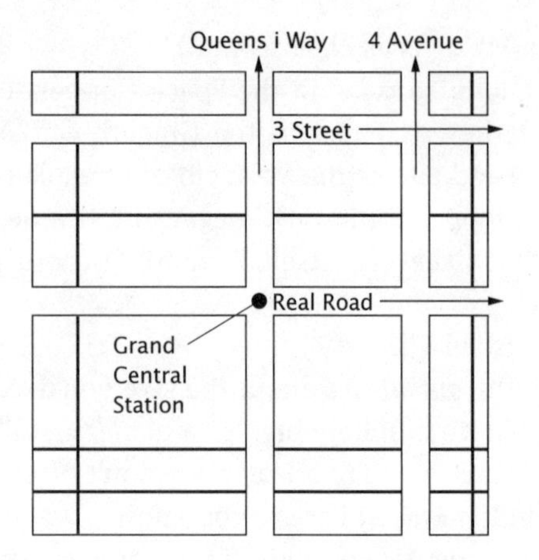

'But I thought the city was infinitely big on the inside?'

'So it is. They're allowed to take infinitely long to escape, you see.'

'But—'

'You'll see in a minute when we look at taxis 2 + 3i and –0.1 + 1i.'

'Those are registration numbers?'

'Very good.'

'They're strange.'

'Very sensible, actually,' said the Mandelblot. 'Each registration number is in the form $A + B$i, where A and B are numbers. And on any journey, taxicab $A + B$i is licensed only to travel from its starting point to whichever intersection is A units to the east (or A units to the west if A is negative), and B units to the north (or B units to the south if B is negative).'

'Let me see if I've got that,' said Vikki. 'Suppose I'm in taxi 2 + 3i. Starting from Grand Central Station, I can make a trip to the intersection of 2 Avenue and 3 Street.'

'That's right,' said the Mandelblot, 'and from there your next trip in taxi 2 + 3i takes you to the intersection of 4 Avenue and 6 Street.'

'OK. And then you can get to the intersection of 6 Avenue and 9 Street, and then the intersection of 8 Avenue and 12 Street, and then the intersection of 10 Avenue and 15 Street, and so on.'

'You got it,' said the Space Hopper. 'The Avenues go up in twos, while the Streets go up in threes.'

'Now,' said the Mandelblot, 'when I said the taxis are trying to *escape* from Quadratic City in infinite time, what I really meant was that they want to follow a path that eventually gets them outside any finite region of the city. An "unbounded" path. So taxi 2 + 3i *does* escape.'

'Hold it!' said Vikki. '*Any* taxi eventually escapes!'

'Except for taxi 0 + 0i. That just gets stuck at Grand Central Station. Which is why my predecessor in this job was just a point. Dot, her name was.'

Vikki found this statement enigmatic, to say the least.

'Look,' the Mandelblot went on, 'the Taxi Controller's job is to make a map of Quadratic City that tells customers arriving at Grand Central Station which taxis can take them out of the city.'

'Nearly all of them,' Vikki pointed out.

'Yes, but someone has to tell them not to jump into the back seat of taxi 0 + 0i. OK, OK, it was an easy job when Dot did it, but not any more. What Dot did was to introduce a systematic way to map out which taxis could or could not escape. Start with a blank map of Quadratic City. Pick a taxi, $A + B$i, say, and colour the intersection of A Avenue and B Street *white* if taxi $A + B$i eventually escapes, but *black* if it doesn't. Do that for every possible taxi, so every point on the map is coloured either black or white. Then all you have to do is look at the map to see which taxis to avoid.'

'If I'm following you, then the only point that gets coloured black is Grand Central Station itself, corresponding to taxi 0 + 0i – the only one that never escapes.'

'Exactly. That's why they called her Dot.'

'Ah,' said the Space Hopper, 'I wondered whether—'

'Another worry!' Vikki butted in. 'Escaping from Quadratic City is all very well, but surely if you've just arrived at Grand Central Station, you might want to see some of the sights before you leave.'

'Mmmmmm . . .' The Mandelblot gave this idea some thought, as if it had never occurred to anyone before. Finally, he said, 'You don't know Quadratic City.'

But Vikki considered this an evasion, and persisted with her line of questioning. 'Suppose I've taken taxi 2 + 3i to 2 Avenue and 3 Street, and I want to go back to Grand Central Station. What do I do?'

'Hop out and hope that taxi $-2 - 3i$ can get to you, of course! Or if that one can't manage it, take a two-trip ride on taxi $-1 - 1.5i$. Use some initiative!'

'That won't work,' Vikki objected. 'Neither of those taxis can ever get to 2 Avenue and 3 Street. Not starting from Grand Central Station.'

'True.' The Mandelblot looked crestfallen. 'I forgot to tell you the other thing that taxis can do, that's why. After Dot had made her map, the city authorities took a look at the result and decided that they needed a more effective taxi service, something that would give the tourists an incentive to stay a while. (God knows why, mind you.) So they decreed that taxis could make a different kind of trip in between their regular "*A* east, *B* north" ones. They're allowed to *square*. In fact, they have to. Each regular trip must be followed by a square trip.'

'Why?'

'No point in following it with a *round* trip, is there? And each square trip must be followed by a regular trip. And they start with a square trip.'

'What's a square trip?'

'Well, that's an interesting question. The easiest way to answer it is to say that if you're on X Avenue and Y Street, then squaring takes you to $X^2 - Y^2$ Avenue and $2XY$ Street. Look, don't blame me, it's the city authorities who thought that one up. See, suppose you're what I call a *real* taxi, one with a license plate like $X + 0i$. So the dot on the map that corresponds to it is where X Avenue meets Real Road, get it? Then when you square, you go to $X^2 + 0i$, which is where X^2 Avenue meets Real Road. That's why it's called *squaring*. It tells you how the Avenues change when you go along Real Road.'

'Oh.'

'*But*, notice that X^2 is always positive, so those intersections all lie to the *east* of Grand Central Station. What if you want to go west?'

'No idea.'

'That's the cunning part – the city authorities weren't totally nuts. Suppose you choose a taxi with license plate $0 + 1i$, that's plain i for short. On the corner of 1 Street and the Queens i Way.'

'Why is it called that?'

'Because points on it correspond to license plates $0 + Yi$, and it doesn't go to Brooklyn.'

'Oh.'

'Anyway, if you work out the square for taxi i you get the intersection of $0^2 - 1^2$ Avenue and $2 \times 0 \times 0$ Street, which is the intersection of –1 Avenue and Real Road. So the square of i is –1, in effect. Every point along the Real Road has a Square Route, and not just the ones to the east of Grand Central Station.'

'That's cool!' said Vikki.

'And at that point I inherited Dot's job, didn't I?' said the Mandelblot. 'And it's a zillion times harder, let me tell you! See, what *I* have to do is map out which taxis escape from Quadratic City if they start at Grand Central Station and keep alternating a square trip with a regular "*A* east, *B* north" trip.'

'And that's hard?'

'Think about it. Suppose you get in taxi –2 + 0i. Where does that take you?'

'Let me work it out . . . we start at Grand Central Station, so the first square trip take us to – Grand Central Station?'

'Yes. The square of 0 + 0i is 0 + 0i.'

'That's silly. Why waste a trip?'

'It puts a few minutes on the clock – kind of entry fee.'

'*Then* the first regular trip takes the taxi to –2 Avenue and 0 Street. Followed by another square, which takes it to 4 Avenue and 0 Street. Then a regular trip . . . ending up at 2 Avenue and 0 Street. And then another square, which gets it to – oh!'

'Problem?'

'It goes back to 4 Avenue and 0 Street!'

'Agreed.'

'And then it just hops for ever between 4 Avenue and 0 Street, and 2 Avenue and 0 Street, doesn't it?'

'Right.'

'So – it never escapes?'

'Exactly. Which means I can take the map, and I can colour –2 Avenue and 0 Street *black*. But you see, that's only one point on the map. There are infinitely many points. But for most of them the calculation doesn't work out so nicely. Try taxi 2 + 3i, since we were talking about that, and see how its itinerary changes with the new rules.'

So Vikki got out her Personal Disorganizer, and programmed its shopping spreadsheet to do the calculation, and this is what it told her:

Trip	*Avenue*	*Street*
START	0	0
Square	0	0
Regular	2	3
Square	–5	12
Regular	–3	15
Square	–216	–90
Regular	–214	–87
Square	38,227	37,236
Regular	38,229	37,239
Square	74,713,320	2,847,219,462

'Looks like it's definitely escaping,' said Vikki. 'I wasn't sure at first, but now that the numbers are getting so big—'

'Yes, but how can you be *certain*?' wailed the Mandelblot. 'You can't keep calculating for ever!'

'Surely,' the Space Hopper protested, 'once the numbers get big enough, the taxi *has* to escape? Can't you prove that?'

'Well, yes, I thought of that,' said the Mandelblot. 'It's a bit complicated, but you can. Turns out that as soon as a taxi gets more than 2 kilometres from Grand Central Station – as the crow flies – it will always escape. But some taxis take an awfully long time to get that far, you see. Here's one of the first taxis I was told to sort out when I started the job. Have a go at taxi –0.1 + 1i.'

Vikki tipped that into her Personal Disorganizer, and got this:

Trip	*Avenue*	*Street*
START	0	0
Square	0	0
Regular	–0.1	1
Square	–0.99	–0.2

Regular	–1.09	0.8
Square	0.5481	–1.744
Regular	0.4481	–0.744
Square	–0.352742	–0.666773
Regular	–0.452742	0.333227
Square	0.0939353	–0.301732

'I don't *think* that one's escaping,' said Vikki. 'The numbers are staying fairly small . . . all of them seem to be within 2 kilometres of Grand Central Station.' She stopped. 'On the other tip, they're not repeating, either, and I doubt they will. Still – I'll guess it never escapes.'

'*WRONG!*' shouted the Mandelblot with glee. 'Keep calculating!'

'Nothing much happening yet – oh, hang on, *now* they're getting quite large . . . and now the numbers are gigantic. So it does escape.'

'You're beginning to see how hard it is to be sure what colour some points on the map ought to be,' said the Mandelblot, 'and sometimes it's a lot worse than that. In fact – ' (he lowered his voice for a moment) – 'the calculation might have to go on for longer than the lifetime of the universe!'

'Which universe?'

'Any universe. As long as you like. Or longer. To tell the truth, whether a general point gets coloured black or white is algorithmically *undecidable*.'

'Which means that no computer program can be guaranteed to give an answer,' added the Space Hopper, helpfully.

'That must have made your job pretty difficult,' said Vikki.

'Difficult? It was impossible! So . . . I *cheated*. But don't tell anybody! I found a rule of thumb that gives a good approximation to the answer, and I used that. Keep following the taxi for a hundred trips: if at any stage it gets further than 2 kilometres from Grand Central Station, then you definitely know to colour the corresponding point white. *Colour all the others black*. Some points will be given the wrong colour if you do that – the ones that were *going* to escape, but slowly. However, only a tiny proportion are like that. And it's just not worth the effort to be more accurate, because the eye wouldn't be able to tell the difference anyway.'

'That sounds a reasonable approach to me,' said Vikki. 'Pragmatic.'

'That's my feeling – but *don't tell my boss*, OK? I might lose my job if she finds out.'

'It's a promise. Not a word shall I breathe. Anyway, what answer did you get?'

'Me. The Mandelblot, I call it, when I want it to sound posh.'

'You mean you're that weird cactus-cat shape because that's the answer?'

'Yes, just as Dot was a dot when *that* was the answer. But I'm far more complicated. In fact, I'm a fractal. Well, my *boundary* is. Look very closely and tell me what you see.'

Vikki cranked up the VUE and zoomed in on the edge of the Mandelblot. 'Sort of cactusy blobs . . . decorated with their own cactusy blobs . . . now it's spirals . . . then sort of – *seahorses*? More spirally things . . . twirly curly curlicues . . . now it's like the branches of a tree . . . Gosh, it *is* complicated, isn't it?'

'Infinitely complicated,' said the Mandelblot, proudly. 'Now, take a look right . . . *here*.'

'It's – my word, it's a tiny copy of you!'

'Perfect, I'm told, in every detail.'

'But the Space Hopper said you're not self-similar.'

'I'm not. Some bits of me – rather rare ones – are small copies of me, though they're all very slightly bent. Most bits, though, aren't. So I'm not made out of small copies of myself. And that means I'm not self-similar.'

'What's your fractal dimension?'

'You mean the fractal dimension of my boundary?'

'Yes.'

'It's exactly 2.'

Vikki was disappointed. 'I expected something like 1.7729 or whatever. A fraction.'

'Ah, but my boundary is a curve. Yet it has the same fractal dimension as a solid disc, or a square. So actually that's very interesting. Because intuitively a curve ought to have dimension 1–'

'It does, in the Henripoincaréan sense,' the Space Hopper interjected.

'Thank you. Anyway, in the fractal sense, the dimension is 2. Which is as big as you can get for a curve in the plane,' the

Mandelblot pointed out proudly. 'So that's *very* remarkable and unusual.'

Vikki turned to her companion. 'Space Hopper, this is fascinating, but is there any serious *point* to it? What's it all about?'

'Dynamics,' said the Space Hopper, 'rules for moving taxis around, if you wish. More to the point, rules for moving *anything* around. Any system that changes as time passes, and isn't subject to random influences, must obey some kind of repetitive *rule* that governs everything it does. That rule is its *dynamic.* If we gave the taxis a different rule for the trips they were permitted to make, we'd get different kinds of dynamic. And almost all important systems change over time. Populations of animals or bacteria, the positions of planets, the weather . . . they're all like the taxis of Quadratic City.'

'So what does the Mandelblot tell us, then?'

'That a very simple mathematical rule can lead to incredibly complicated behaviour,' replied the Space Hopper. 'Isn't that amazing?'

'I guess . . . But doesn't that mean that trying to understand dynamics in terms of rules is a waste of time?'

'Not at all!' said the Mandelblot heatedly. 'Can't you see how beautiful I am?'

'Sorry, but what's beauty got to do with anything—'

'Pattern!' shouted the Mandelblot. 'Structure! I may be infinitely complicated, but I'm made from layer upon layer of intricate pattern!'

'That's right,' said the Space Hopper. 'What the Mandelblot tells us is that something which seems very complicated may in fact arise from simple rules. So the trick is to understand the rules, not the complicated behaviour that follows from the rules. And if you didn't know there *were* any rules for the Mandelblot, you'd still be able to tell there must be some, because of the patterns. He's not a random mess, you see.'

'I'm not any kind of mess,' protested the Mandelblot.

'Sorry, never meant to suggest you were,' said the Space Hopper. 'What I mean is, instead of all this telling us that rules aren't any use because they can lead to complexity, it tells us that we can hope to understand complexity by finding out what the rules are.

But, to do that we have to be sensitive to new kinds of pattern, such as fractals. New types of pattern can lead us to new types of rule.'

Vikki felt she *ought* to be impressed by this, but wasn't sure exactly why. The dilemma must have showed on her face, because the Space Hopper tried again. 'The Planiturthians have a name for rules like that, Vikki. It may be an exaggeration, but it shows how important they can be.'

'And what name is that, Hopper?'

'Laws of Nature.'

'Heavy,' said Vikki. 'Philosophical. And all based on a city that's so featureless, the only thing new arrivals can think about doing is getting out again.'

'Well, if you feel like that, we'd better go and visit somewhere else.'

That sounded a good idea to Vikki. 'Somewhere *fun*, Hopper? Where there's a lot going on?' *Somewhere to take my mind off home.*

The Space Hopper wiggled its antennae, searching for inspiration. Then its face lit up in a ∪. 'I know just the place!'

THE TOPOLOGIST'S TEA-PARTY

An instant – or a lifetime – passed. Vikki had no way of telling how much time had elapsed, since time had no meaning in metaspace . . . What she experienced was not a transition, but a shift of attention. Her Virtual Unreality Engine reconfigured itself, and she was . . . *elsewhere.*

Using her VUE-enhanced senses, she tried to come to grips with her new surroundings. It was difficult, for nothing stayed still. Strange bendy plastic shapes drifted silently past, shrinking and swelling, rippling and undulating, twisting into grotesque pretzel-like forms, writhing and intertwining like very friendly surreal baroque worms. Even the ground beneath her endpoints heaved and pulsated like a living creature, sprouting elaborate protrusions and then sucking them back into its weird alien landscape. It was disturbing, yet oddly beautiful.

'Here we are,' said the Space Hopper proudly.

'Where?'

'Topologica,' replied the Space Hopper, 'the Rubber-sheet Continent, which doesn't so much drift as *stretch* . . . We have entered the realm of topology, from which rigidity was long ago banished and only continuity holds sway. The land of topological transformations, which can bend-and-stretch-and-compress-and-distort-and-deform' (he said this all in one breath) 'but not tear or break. You'll find plenty to occupy you here. Look, someone's coming to meet us already.'

Vikki was about to ask what such a place might be good for when she looked up to see a small, rather nondescript creature sauntering towards them through the ever-changing hills and valleys of the restless landscape. It carried some kind of bag. Even though the creature seemed to be dawdling, it reached them surprisingly quickly.

And promptly went to sleep.

Vikki stared at the snoring animal. It had a fat little body, a sharp nose with whiskers, tiny ears, and a long thin tail. She poked it, being careful to fold her tip back on itself like a hairpin to avoid spiking the curious beast. It continued to snore, and where she poked it her curled-up tip left a dent. The animal had a texture like dough.

'Meet the Doughmouse,' the Space Hopper whispered. Then he shook his horns in irritation. 'I don't know why I'm whispering. A Doughmouse can sleep through just about anything . . .' his grin broadened into a ∪ '. . . except a "***WAKE UP!***" . . .' the Doughmouse shot to its feet and the contents of the bag it was carrying scattered everywhere '. . . call from a Space Hopper. Doughmouse, you have company.'

'Oh, hi . . .' The Doughmouse gave a gaping yawn, and began to curl up again.

'I said, ***WAKE UP!*** Be polite this time, and *stay* awake.'

'I'll dough my best,' said the sleepy creature.

'Hopper, you've made the poor little thing spill everything,' said Vikki, her latent maternal instincts aroused by the animal's big round eyes and doleful expression. 'Let me help you pick them up.' The bag was marked 'Doughnut Disturb', and several dozen ring-shaped tubes of cooked dough had spilled out onto the ground.

Moments later, they had contorted themselves into the cups from a twelve-piece tea-set. Chinese willow pattern.

'Doughmouse,' said the Space Hopper, 'allow me to introduce my friend Victoria Line.'

'Dilated to meet you,' said the Doughmouse, swelling alarmingly as if to prove it. Vikki's maternal instincts were about to vanish, but then it shrank back to its previous size. 'Er . . .' The Space Hopper had told it to be polite, and the Doughmouse was a very literal-minded creature. What was the height of politeness when meeting a stranger? Ah, yes! It turned to Vikki and bowed. 'Would both of you care to join me—'

'Join you to what?' asked the Space Hopper. 'That's not a topological transformation!'

'—for tea?'

'That's very kind of you,' replied Vikki, ignoring the byplay, 'but your tea-set has only cups, Mr Doughmouse. No saucers, no plates – and no teapot! I'd say it's a very poor service.'

'And that's a very poor joke,' said the Doughmouse, 'also a very old one. And you're a very forward young lady. Only just arrived, and already you're criticizing my crockery.'

'Vikki is a visitor from Flatland,' the Space Hopper said.

'Oh well, that explains it,' the Doughmouse stated with a shrug. 'Very elegant people, Flatties, but rather rigid in both their geometry and their thinking. And yet . . . here you are, Victoria, consorting with me and the Space Hopper, so perhaps you've got more imagination than your fellow geometrymen. I wonder, though, if you have enough imagination for *this*! You criticized my tea-set, and here is my answer!' And with that, the Doughmouse tipped out two dozen ball-shaped doughnuts. Then he turned the bag upside down and jiggled it, muttering to himself, until the bag disgorged the most peculiar doughnut Vikki had ever seen.

It had two holes.

The Doughmouse looked expectantly at her.

After a few seconds' stunned silence, she realized she was expected to react. 'Sorry, but I don't . . . what I see is a dozen cups and 25 doughnuts, of which 24 have no hole at all and the other is so very badly made that it's got two. It still isn't a tea-set!'

The Doughmouse gave the Space Hopper a meaningful look. 'It is if you think like a topologist. Let me present my wares in a way that makes sense to your Euclidean eyes.' He held up one of the round doughnuts, and as she watched it flattened itself into a thin pancake, for all the world like a lump of pastry being rolled flat. The pancake's rim tilted and its edge contracted. With a final silent hiccup, it settled into its new shape. The Doughmouse placed the now-transformed doughnut beneath one of the cups.

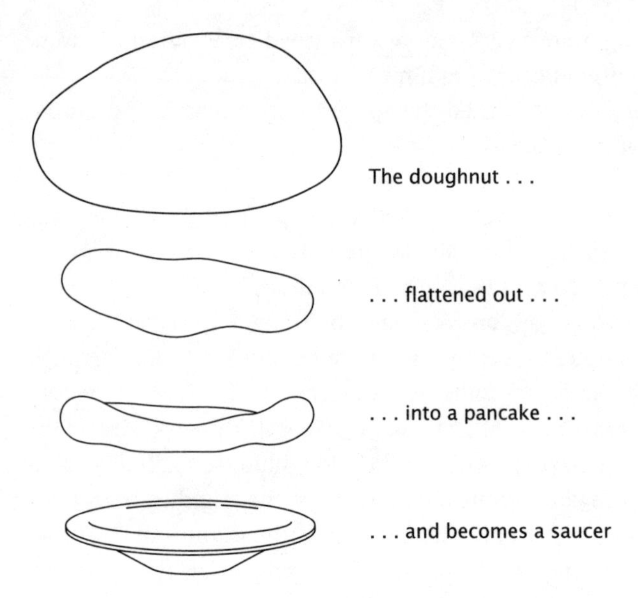

'A saucer!' cried Vikki.

Eleven more doughnuts changed into saucers. Twelve more expanded to about twice their size and became plates. Only the two-holed doughnut remained doughnutty.

'Well . . .' said Vikki, half-conceding. But not yet. 'Where's the teapot?'

The Doughmouse used its whiskers to gesture towards the two-holed doughnut.

'That? *That's* not a teapot! Teapots don't have two holes!'

'Really?' replied the Doughmouse.

'No. The tea would trickle out.'

'Oh dear, just my misfortune to bump into a teapot expert! Well, I'm sure you know best. But tell me: does a teapot have a handle?'

'Of course. Otherwise you couldn't pick it up,' said Vikki, getting angry now. The Doughmouse was so *obtuse.*

'So there must be a hole between the pot and the handle for you to put your hand through?'

'Hand?'

'Sorry, I forgot you're a Flatty. Your manipulatory vertex.'

Vikki hadn't thought of the aperture as a hole before, but she had to admit that it was just as much a hole as the one in a doughnut – which wasn't actually *in* the doughnut . . . it was more a place where there could have been dough . . . but . . . there wasn't.

'All right, I accept that in that sense a teapot has one hole. But your doughnut has two . . .' Her voice trailed off. 'Oh! A teapot has a hole in the top!'

The Doughmouse gave her an enquiring stare. '*That's* not a hole.'

'Yes it is! It's a hole for putting the tea into the pot.'

'Ah, but that "hole" only goes *in*. It doesn't come out again.'

'Nonsense!' said Vikki. 'A hole in a wall can go in but come out again. A hole that goes in and does come out again is called a *tunnel*.'

'Terminology, terminology,' muttered the Doughmouse, 'why does everybody get hung up on *terminology*? It's enough to send a body to sleep . . .' It emitted a tentative snorelet, and the Space Hopper administered an admonitory kick. 'Ow! Uh – in Topologica, a hole that only goes *in* isn't a hole at all. It has to come *out* again. Somewhere else.'

'Then why didn't you say what you meant?'

'I did say what I meant – you just didn't *hear* what I meant . . . but don't get me started on the meaning of words, young lady, it's been done. Tell me, when you pour the tea, how does it get out of the pot?'

'Through the spout,' said Vikki. Really, the creature was *extremely* annoying.

'So there has to be a hole in the spout to let the tea out.'

'That's a tube, not a hole!'

'A tube,' said the creature carefully, 'is just a long, thin hole.' It coughed in a self-important way. 'In Topologica, the Rubber-sheet Continent, "long" and "thin" have no significance. A hole is a hole, whatever its size or shape. Though I should be careful how I say that. What do you make of *this* doughnut?' It held up a semitransparent, strangely convoluted mass of dough:

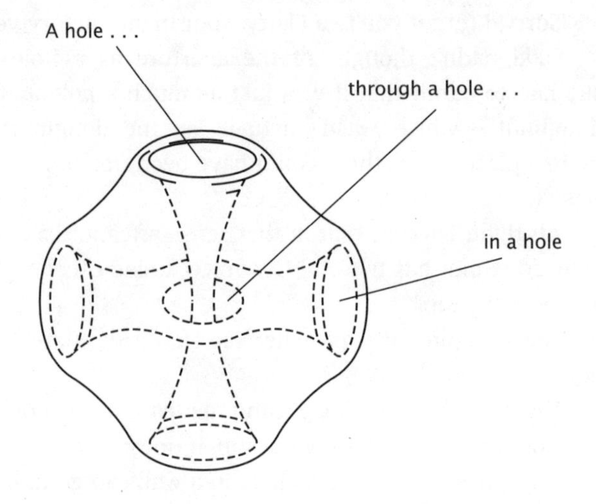

'It's a . . . no, it's more of a . . .'

'It's a hole-through-a-hole-in-a-hole,' proclaimed the Doughmouse.

'That doesn't make a hole lot of sense—' Vikki began, and then stopped. Because actually, it did. Suddenly everything that had been happening *was* starting to make a weird kind of sense. The Doughmouse wasn't as stupid or as annoying as he seemed. 'Hole' was just a rough-and-ready way to describe something infinitely more subtle: *topology*.

'You're telling me that the hole where the tea goes into the pot, and the spout where it comes out, are really just the two ends of the *same* hole?'

'Exactly! So now you can see how my two-holed doughnut turns into a teapot. Think of it as two doughnuts joined together, like Siamese twins. First, I pinch one twin down to make a handle on the side of the other. Then I flatten the bigger one out to make a disc with a small hole. I bend the disc up into a cup-shape, and then stretch it further and shrink the edge of the cup so that it's teapot-shaped. Finally, I pull out the region immediately around the hole into a tube, stretch that to the right length, and bend it into a spout shape – oh, and while I'm at it, I tidy up the shape and position of the handle. Done!'

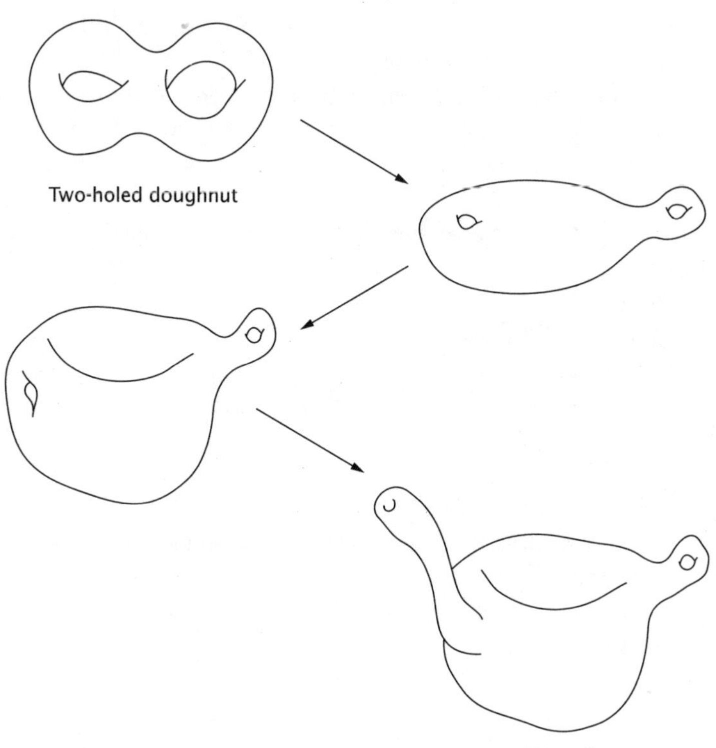

Topologica, thought Victoria, *is a* very *strange continent. I must be careful. Things aren't necessarily what they seem. I wonder if* – she picked up one of the cups and nibbled gingerly at the rim.

Sweet dough crumbled into her mouth.

She looked inside the teapot. It was full of tea, and none of it seemed to be leaking out. She poured some tea into her half-eaten cup, expecting the cup to soak it up and turn into a sloshy mess. It didn't. It sat there, like normal tea in a normal cup.

'All geometric forms in Topologica lead multiple existences', said the Space Hopper, who had guessed what she was worrying about. 'Mathematically, any shape is topologically identical to anything you can continuously deform it into. But to the inhabitants of this particular world, the *physical* properties of shapes depend on what you use them for. "Context-dependent", that's the phrase.'

'Oh.' Vikki began to pay more attention to the tea, rather than the cup. 'Yuck! There's no milk!'

'An omission that can easily be remedied,' said the Doughmouse. 'I happen to know that just over the brow of the next hill there lives a cow who will sell us a bottle of milk at a very reasonable price.'

'What is the name of this beneficent beast?' asked the Space Hopper.

'Moobius.'

'And its milk comes in . . .?'

'Klein bottles.'

The Space Hopper sighed. 'I rather thought so.'

•

Wunday 9 Noctember 2099

Whatever next, Diary Dear? A talking cow, would you believe? Not that it *looked* much like a cow – but, after all, a doughnut doesn't actually *look* much like a teacup. Even so, this was a very singular cow . . .

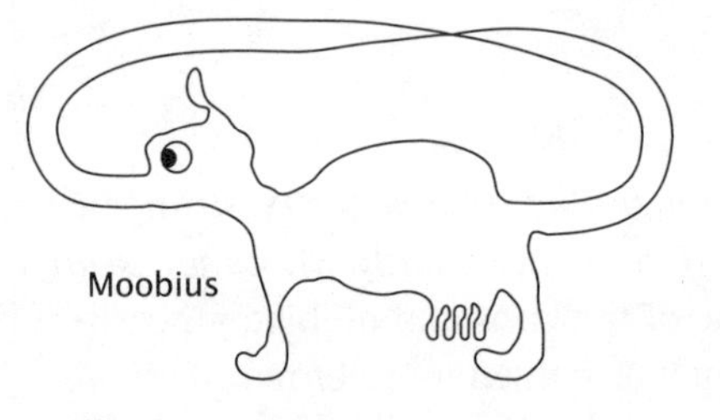

•

Victoria wasn't particularly bothered that Moobius the cow was two-dimensional – because, after all, so was she. Nor was she bothered by its extraordinarily long tail, which could wrap all the way round to touch its face. She was a little disturbed, though, when she realized that the cow's tail was glued to its nose. Or, rather, that tail and nose merged seamlessly into each other:

Was the creature *really* a cow with its tail attached to its nose? Might it instead be an elephant with its trunk attached to its bottom? Either way, she wasn't sure she wished to pursue the matter. However, even the remarkable appendage, whatever it was, was less disturbing than the extremely loud music that seemed to be coming from somewhere inside the cow. They had heard the music growing ever louder as they approached the animal, but for a while they had assumed it must be coming from somewhere else. Now it was clear that the music definitely originated inside (or, the creature being two-dimensional, *on*) the cow itself. At first it had seemed to be coming from the head, but then it migrated to the stomach. At the moment it seemed to be concentrated near the cow's rear end. It was some kind of military march – mostly brass and drums.

'Hello!' shouted Vikki, trying to make herself heard over the din.

'Eh?' said the cow.

'Sorry, but your music is making an awful lot of noise!'

'You'll have to speak up,' said Moobius, 'my music is making an awful lot of noise.'

Vikki put her moth close to the cow's ear and yelled, 'Can't you put a stop to it?'

'Good heavens, no,' said the cow. 'That would be terribly orienting.'

'Surely you mean "disorienting"?'

Moobius shook its head. 'I know what I mean.'

'Well, can you turn it down a bit so that we don't have to shout all the time?'

'I suppose so,' said the cow, somewhat miffed. It jiggled around for a moment, and the noise of the band faded to a low *clickety-clackety-boom*. 'I'd be nothing without my band,' said the cow. 'It's essential to my existence.'

Vikki decided that this was yet another aspect of Moobius the cow that might best be left unexplored, but the Doughmouse must have felt otherwise. 'Oh no it isn't,' it said.

'Oh yes it is!' the cow responded, quick as a flash.

'Oh no it isn't!' Vikki felt that the quality of debate might be improved, but there was no doubting the level of commitment.

'Oh no it isn't!'

'Oh yes it is!'

'Oh yes it is!'

'Oh no it isn't!'

'Oh no it isn't!'

'Oh yes it—'

'*Hold it right there,*' Vikki interrupted, replaying the conversation in her head 'I don't think I *quite* followed that—'

'Let me explain,' said Moobius in a patronizing tone of voice. 'First *he* said "Oh no it isn't" and then *I* said "Oh yes it is" and then *he* said—'

'Not at all,' the Doughmouse objected, '*you* started the argument by saying that it was, though not in those words, and then I made the reasonable point that it wasn't, to which you responded without a single *whit* of justification that it most assuredly *was*, and then—'

'*Eeeeeeeeek!*' Vikki's scream stopped them both. She gathered her breath and tried to calm herself. 'Uh . . . Doughmouse: first you started out saying that Moobius's band *isn't* essential to its existence.' She turned to appeal to the Space Hopper for support, but it seemed to have wandered off. 'Uh . . . when Moobius disagreed, you repeated what you'd said before. But then . . . well, Moobius changed his mind and *agreed* with you, but you promptly changed sides and said that the band *is* essential to Moobius's existence. Then Moobius agreed with *that*, too, but all *you* did was go back to your original position!'

'Sure. But then Moobius changed his mind again—'

'*And so did you!* You were both being completely inconsistent! Don't either of you know which side you're on?'

They stared at her. 'I know which side *I'm* on', said the Doughmouse.

'Which?' demanded Vikki.

'The inside,' said the Doughmouse.

'The *what* side?'

'And you are on the outside,' the Doughmouse added in a helpful tone.

'Uh . . . then what side is Moobius on?'

'Both,' said the cow. 'Well, actually there's only one, and it's on me.'

'You're both talking utter rubbish,' said Vikki.

'Not at all,' said the Doughmouse. 'Haven't you heard the phrase "a side of beef"?'

'Yes, but—'

'Well, that's what Moobius is. A side of beef. Whereas I—'

'Yes?'

'—am an inside of a Doughmouse. And you—'

'*What about me?*'

'—are nearly beside yourself with rage. And I recommend that it remains "nearly", my dear, at least until you visit a space with the property of bilocation.'

With an effort, Vikki took a deep breath before trying to reboot the conversation. 'Moobius. Is. An. Entire. Cow. A side of beef is half a cow, that's why it's called a *side*. If you slice a cow in half from nose to tail – excuse me, Moobius, nothing personal – you get *two* separate sides of beef. One from the left side and the other from the right side. Even Flatland butchers know that, though our cows are mostly oxagons, you know. Er – a cross-breed between hexagons and octagons,' she added apologetically.

'No hoctagons?'

'Well, no, but there's a hogsagon, which yields excellent bacagon—'

'*Keep to the point!*' shouted Moobius. 'I assure you that I am a *single* side of beef.'

'Rubbish!' said Vikki. 'Nothing has only one side.'

'Well, actually, *nothing* has no sides at all—'

'And *you* were telling the Doughmouse to keep to the point! Of course you've got two sides. Why, even from here I can see that part of you is on the far side and the rest is on the near side.'

The cow grimaced. 'Yes, I *know* that, you pigheaded Flatty, but—'

Vikki made an effort and kept her temper. 'So you've got two sides, Moobius.'

'Only . . . locally. Um . . . you know, what I could really do with is a hosedown. Can you fetch that hose over there, Doughmouse? And since the young lady is so clever and knows so much about bovine anatomy, she can hose me down on *just one side*. Yes?'

'Well – I really don't see why *I* need to do it, but in order to prove my point, OK.'

The Doughmouse came back with a dog collar. 'Got it.'

'No, that's a—'

'In Topologica, nothing is what it seems.'

'No,' said Moobius, 'nothing *isn't* what it seems, because if it's nothing then there isn't anything to *seem*.'

'Oh, do shut up!' snapped the Doughmouse. He handed Vikki the dog collar. 'Grab one end, Victoria.'

'It hasn't got any ends.'

'Yes it has, they look like edges. Grab an edge.' And she did . . . and the Doughmouse pulled . . . and pulled . . . and pulled . . . and the short fat cylinder of the dog collar became longer and longer and longer, and thinner and thinner and thinner, until she found herself holding the end of a hose. She just managed to point it away from herself before a spray of water shot out.

'Hose me! Hose me!' yelled the cow. 'But only on one side, remember!'

'Easy!' said Vikki. First, she peeped round the back to make sure that there was a second side to the cow's face: there was. Then she sprayed water all over the side of the face that was nearest to her. Slowly she worked her way along the flank of the cow until she reached the tail.

'Carry on,' said the cow. 'You haven't finished yet.'

'I know that. But there's only the tail left to do.' Vikki worked her way back along the cow's tail, checking that the near side was dripping wet but the far side was dry, but about halfway along she found she was having to contort her body to get at the correct side of the tail. 'Hey, Moobius?' she called out. 'Your tail's twisted, did you know that?'

'Tell me about it!' said the cow.

The Doughmouse started to yawn again, and looked round in a vague sort of way for something soft to sleep on. It took the teapot out of its bag, did something to make it grow a good deal bigger, climbed in, and pulled the lid over its head.

Vikki kept on spraying water over Moobius, being careful to keep the far side dry. 'Nearly done, coming back to the nose – hold it.'

'Now what?'

'I'm back to your nose – but it's *dry*!' She glared at it. 'Moobius, have you been sneakily drying yourself while I've been hosing your tail?'

'Not a bit of it,' protested the cow. 'Try looking on the other side.'

'It's – it's dripping wet! Are you playing a trick on me?'

Moobius nodded. 'Yes.' Then it shook its head. 'But not in the way you mean.'

'Can't you ever make up your mind?'

Moobius gave her a level look. 'I always thought that my mind made *itself* up, and only told *me* about it later. But the wet "side" of my face really is the one you started washing, I promise.'

'Then what – *oh*. The twist. In your tail. It – it flips over so that the wet side runs into the dry side. I see it all now. If I keep going I'll spray the whole of your *other* flank, and then go along the *other* side of your tail . . . and finally get back to where I started. The twist makes your two sides join together.'

Moobius nodded. 'Which means that what you thought were two different sides are actually just different parts of the *same* side. So, like I kept telling you, I've only got one side. And I'm not on it: it's on me.'

'Whereas I—'

'Have two separate sides,' said the Space Hopper, reappearing from nowhere with two new companions. 'These guys were looking for the Doughmouse – Vikki, where'd he go?' A resonant snore boomed from the interior of the teapot. 'Oh. You shouldn't have let him go in there. Now we'll have a terrible time waking him up.

'Anyway, *you* have two separate sides because you've got edges that keep your sides separate.'

Vikki thought about this. True, an edge can separate one side of a flat surface from the other, but the Space Hopper's argument had a flaw. 'Moobius has edges, too.'

'*Edge*,' the Space Hopper corrected. 'Run your eye along it. What you think ought to be two separate edges *also* join up seamlessly.'

'Because of the twist. Again.'

'Of course. Every good tail has to have a twist, you know.'

'Um – I think that's "tale"', said Vikki.

'Tail, tale, so what? It's the *twist* that matters—'

'—so you'll only get one side of beef out of me', Moobius finished.

•

Wunday 9 Noctember 2099 [continued]

. . . and that, Diary Dear, is how two people can hold a really interactive but one-sided conversation.

Moobius went on to say that the reason we were getting so confused is that the word 'side' has several meanings. It's not really a question of how many sides a cow has: it's *orientations* that matter.

I'll see if I can explain. Moobius has a resident band – mostly brass, but when I looked closely I saw that there's a drummer with a gigantic bass drum. It always carries the drum on its left – well, what Moobius says is its left. When I first saw it, that's where the drum was.

But when the band marched all the way along Moobius's body, back along its tail, and returned to its starting-point, the drum was on the *right*.

I'd been watching it closely. The drum never switched sides.

Then it clicked. The band doesn't live *on* Moobius. It's *in* it. Like an ink drawing that's soaked right through to the other side. Now, *locally* – that is, near any given place – Moobius has two sides, just like any other surface. But globally – taken as a whole – the twist in its tail makes the two sides join up. As a result, there's no sensible concept of left/right. As the drummer goes round the twist – Diary, stop laughing, you know what I mean – his apparent orientation gets flopped over, so left becomes right and right becomes left.

The Moobius band is nonorientable, that's the point.

And Moobius isn't the only such surface. There's an even weirder nonorientable surface, which Moobius keeps her milk in. A Klein bottle.

The Space Hopper told me why it's called that. It's a Planiturthian name, you see. A mathematician's *joke*. In one region of Planiturth, called Germany – no, not because it has many germs, Diary Dear – the word *klein* means 'small'. But size doesn't matter in Topologica, so that's a red herring. And a Germany-People called Felixklein invented a nonorientable surface in 1882. Unlike Moobius, it had no edges – it was a *closed* surface. Basically, it's a tube that bends round and joins up to itself, like a doughnut.

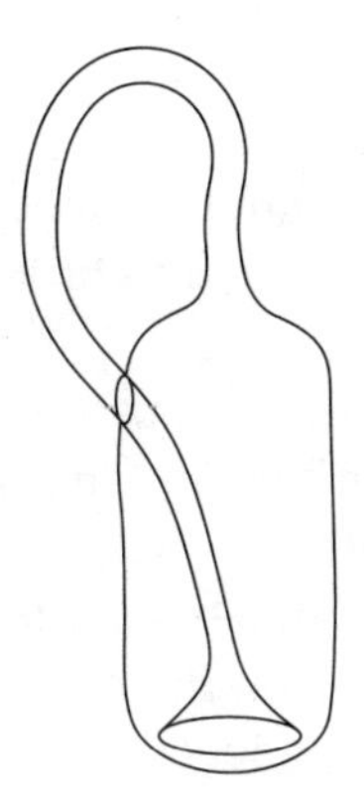

But as it bends round it kind of passes through itself and turns inside-out, so the surface all joins up.
Weirder still, it only passes through itself if you try to embed it in 3D. In 4D it *doesn't.*

Now, the Space Hopper has a private theory about the name *Klein bottle.* He thinks it was originally *Klein's surface.* In Felixklein's day, Germany-People were forever inventing new surfaces and getting them named after themselves. For instance, there was a People called Ernstkummer and he invented *Kummer's surface* – which in Germanyspeak is *Kummersche Flache*, the '-sche' being a possessive ending and *Flache* meaning 'surface'.

So, the Space Hopper reckons, the name started out as Kleinsche *Flache*, Klein's *surface.* But it *looks* like a bottle, and the German for bottle is *Flasche*, so . . .

QED, as we Flatlanders say.

•

The two new creatures the Space Hopper had brought with him had come to join the tea-party. Make it more sociable, he said. Vikki thought this unlikely, though she politely kept the thought to herself. One of the creatures was a very angry-looking horse. The other carried what seemed to be a bottomless bucket of mud, and kept building mud huts with it.

'Who are they?' whispered Vikki.

'The Harsh Mare and the Mud Hutter,' replied the Space Hopper. The two creatures climbed into the teapot to join the Doughmouse, waking it up, and all three began to bicker. Their voices rose and fell.

'I know they *seem* a bit intimidating,' the Space Hopper continued, 'but that's because they *are* intimidating, so it's OK to be intimidated by them, right?'

'But I don't want to be intim— *What the convex hull is that?*'

Another extraordinary inhabitant of Topologica was trotting towards them across the bouncy surface of the Rubber-sheet Continent. Its head, if head it was, was adorned with antlers, if antlers they were . . . they seemed to branch and entwine and

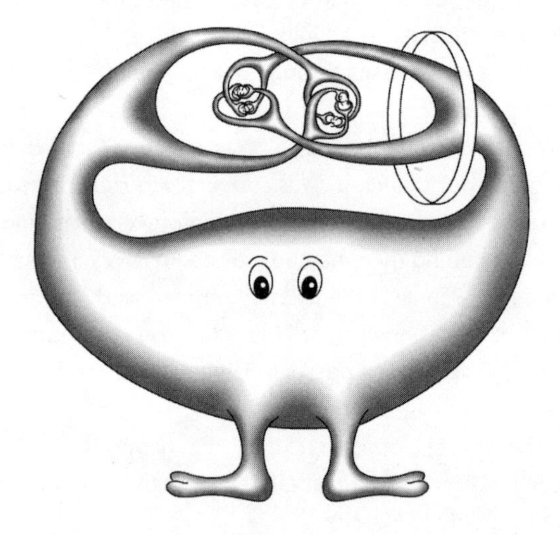

branch and entwine . . . and the closer the curious beast got to them, the finer the branches became.

A red sweatband was wound between them.

'Wow!' said the creature, skidding to a halt. 'A tea-party? Great! And visitors from Out-a Space, too! Hi, Hopper – how's tricks?'

'An old friend,' the Space Hopper whispered by way of explanation. 'Whatever you do, don't take him seriously.' He turned to the new arrival and spoke in a more normal tone. 'Vikki: meet my old buddy Alexander the horned sphere.'

'You're a sphere?' Vikki found this difficult to believe. Alexander surely wasn't *the* Sphere: Albert would undoubtedly have mentioned the horns.

'Hey, baby, yeah!' boomed Alexander.

'Er, in a manner of speaking,' the Space Hopper qualified.

'Yeah. My inside is topologically the same as the inside of a Sphere – but my outside is topologically different from the outside of a Sphere. Dig?'

'That doesn't make any sense at all,' said Vikki.

'Sorry to contradict, babe, but it does. You're forgetting the difference between the intrinsic topology of a space, considered on its own, and what happens when it's embedded in another space. Intrinsically I'm a Sphere, but extrinsically I'm not.

'It's the horns, see?' Alexander went on. 'From the inside, pushing out a horn doesn't make me any less spherical. So pushing out lots of horns, even infinitely many, doesn't either. But when seen from the outside, the horns tangle up, whereas the outside of a sphere isn't tangled at all. I'll prove it. Try to remove my sweatband.'

'Uh . . .' Vikki began. Those horns looked *sharp*. Fractal, even.

'Go on, try,' the Space Hopper encouraged her, 'give it a tug.'

Vikki tentatively pulled on the sweatband, but she couldn't dislodge it from its loop around the horns. 'It's . . . stuck!'

'"Homotopically nontrivial" is the phrase ya want, babe, but ya dead right. But . . .' The horned sphere's horns started to shrink back into its body. 'If I now turn myself into a bog-standard sphere . . . Try again, kiddo.'

By now the horned sphere had turned into a perfectly round ball. The sweatband was wrapped round it somewhere close to what a Planiturthian would have called the Tropic of Cancer.

Vikki reached over and, without the slightest difficulty, lifted it off.

'See? My *inside* topological shape never changed, but now I've got a null-homotopic sweatband!'

'Uh?'

'It can be pulled off, OK? So my outside topology *has* changed.'

'Indisputably', said Vikki.

The bickering inside the teapot stopped, and the Harsh Mare

poked her nose over the edge. '*You* may not dispute it, madam, but *I* will dispute *anything*!'

'No you won't,' came the voice of the Mud Hutter.

'Yes I will,' insisted the Harsh Mare, ducking back into the teapot. *Snore*, the Doughmouse contributed to the debate. The bickering intensified.

'Oh dear,' said the Space Hopper, 'I think we may have outstayed our welcome. Time we went.'

And they did.

7

ALONG THE LOOKING-GLASS . . .

'This stuff about one-sided surfaces is all very interesting,' said Vikki, 'but it's just intellectual fun and games, isn't it? You can't really *use* it for anything.'

The Space Hopper gave a few thoughtful bounces. 'That depends on what you mean by "use",' he said. 'If you mean *direct* application, then maybe not. I did read once about someone using a Moobius band as a conveyor belt, because that way the wear on its surface was halved – but you could achieve the same thing by using an ordinary cylindrical belt and then turning it over to use the other side, so I can't personally see the point. Except perhaps a small saving on maintenance.'

'So you don't—'

'I don't think there are many *direct* applications. But most applications of maths are indirect. It plays out its roles behind the scenes. And there are plenty of ways to use topology. But you were talking specifically about one-sided surfaces. I *do* know one place where such things turn up naturally, and the associated ideas are certainly useful. Ready? Let's go!'

Vikki scarcely had time to prepare herself for the transition before her VUE reconfigured . . .

•

This was a bleak and oppressive world. A seemingly endless sandy desert stretched away in all directions. It was flat and featureless.

'I don't like this place,' said Vikki.

'Why not?'

'It feels lonely and unforgiving. And nothing's happening.'

'Not yet. Don't judge geometries by first appearances, Vikki. Some of the most beautiful geometries start out looking most austere. If you reserve judgement, you will find that this is a very elegant world indeed.'

'Hmmph.' Vikki was unimpressed. 'Where are we, anyway?'

'We have landed somewhere on the Projective Plain. And it's not as deserted as it seems. It's teeming with wildlife.'

'Wildlife? There isn't so much as an *ant*!'

'Not so. The Projective Plain is inhabited by lions.'

'Lions?'

'Projective lions.' The Space Hopper saw Vikki gasp, and her eyes flashed wildly to and fro. 'Don't worry – they're friendly. So friendly, in fact, that *any* two projective lions always meet.'

'Meet? Meet what?'

'Each other.'

'Meet each – *where?*'

'Somewhere. At some projective point – which is to say, at some location in the Projective Plain. Precisely where depends on the lions. Why, right now a lot of them are all meeting *here*.'

'Lions? Lots? Here?' Vikki's voice was edging upwards.

'Don't worry, they're harmless. Anyway, you're a lion yourself.'

'Pardon?'

'Victoria Lion.'

'That's Victoria *Line*, you buffoon!'

'Which, in the local jargon, is pronounced "lion". The difference is that *you* are a Euclidean lion – well, part of one, anyway – whereas *they* are projective lions. And it's a very significant difference, because in Flatland you can have parallel lions – sorry, *lines* – which never meet, but on the Projective Plain every lion meets every other lion at one, and only one, point.'

'Crazy!'

'Not at all. And to get back to your original question: just like a Klein bottle, the Projective Plain has only one side. And it's very useful, for example in art and perspective drawing.'

'To me it looks like an entirely normal plain – I mean, plane,' said Vikki.

'Yes, it does, and *looks* is definitely the word. But as we explore the Projective Plain and get to know the lions, you'll come to see that I'm telling the truth. This is a far stranger world than it appears.'

'It certainly is,' said Vikki. 'For a start, I can't see any lions.'

'You can't? Oh, my mistake – I forgot to adjust your point of VUE to make them visible. They are infinitely thin, you see. Well, they would be, wouldn't they?'

'Why?'

'Because they live in a desert and they're very friendly. So of course they don't get much to eat. And that makes them infinitely thin, so to the untutored eye they are completely invisible. But if I make the right adjustments to the VUE's settings . . .'

Suddenly a blue glow came into being – a long, thin line, stretching away to the distant horizon. Vikki turned, and it stretched away behind her to the horizon in *that* direction, too.

'That's a projective lion?'

'Apart from the artificial blue tinge, yes.'

'It looks just like an ordinary lion to me. I mean, "line".'

'It does, but it's not. Let me reveal another lion, and you'll see the difference immediately.' Now a thin red lion brightened into visibility.

Vikki stared at the two lions. They marched off to the horizon together. As far as she could see, they were perfect parallels. But the Space Hopper had said . . .

'I thought you said there were no parallel lions in the Projective Plain?'

'I did.'

'But these two lions *are* parallel!'

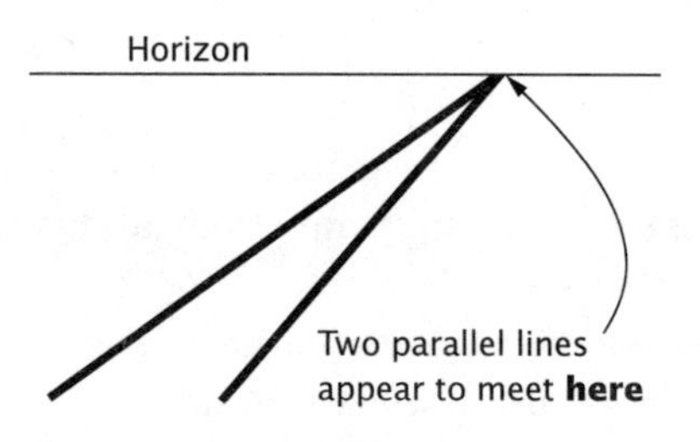

'Really?'

'They don't meet anywhere!'

The Space Hopper did a double take. 'They don't?'

'No. They seem to converge as they go towards the far horizon, but that's an optical illusion.'

'It is?'

'Well, it would be on Flatland. I'd have thought that on the Projective Plain . . . oh, maybe not. New space, new rules.'

'Now you're talking. This space runs on projective geometry, not Euclidean geometry. Think how many paintings include the horizon as a genuine line. Why is that? One reason is that when artists want to paint a scene, they *project* that scene onto a flat sheet of canvas or paper. Another is that Planiturth is round, but even if it were an infinite Euclidean plane, they'd still draw a horizon – because if you looked, that's what you'd see. In effect, they draw lines from every object in the scene to their eye, and where that line hits the canvas, that's where they paint that object. It's like looking out of a window and tracing what you see on the glass. If the object is at infinity, in the projective sense, it may still correspond to a point on the glass – and if it does, you have to draw it to make the picture look right. So projective geometry is a good geometry for thinking about perspective.'

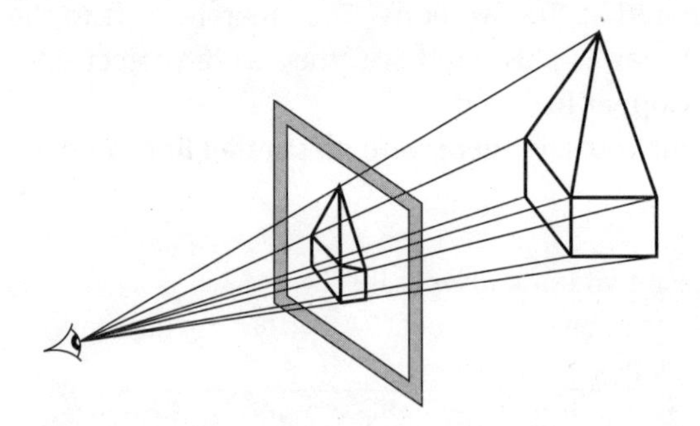

Vikki was assailed by an unexpected twinge of pain. 'Damn, I'm getting blisters on my endpoints from all this space travel. I'm going to take my shoes off.'

(You may be wondering why a line needs shoes – and if so, where she wears them. Well, shoes are a universal female attribute throughout the unbounded Mathiverse. They're an essential fashion accessory. Moreover, they always come in pairs – otherwise they would not be shoes. To be sure, the millipods of the Amazin' Basin have to wear five hundred pairs, but that's an extreme case. Victoria Line wore a normal pair of shoes: a left shoe for her front endpoint, and a right shoe for her rear one. She was always getting them mixed up, and that was a nuisance because then they didn't fit.)

She eased off her shoes and inspected her endpoints for blisters. The left end was fine, so she put that shoe back. But the right end was sore.

The Space Hopper made vague passes in the air, and a container of ointment appeared. He passed it to Vikki, and she massaged some into her sore foot.

'Let me ask you something, then,' said the Space Hopper, as if their conversation hadn't been interrupted. 'Is the horizon a genuine *line* in Flatland?'

'Yes.'

'Then where is it?'

'A long way away.'

'How far? If the horizon exists, then it must be in a definite place. So how far away is the point on the horizon where two parallel lines would meet – if they were extended far enough?'

'Well . . . at the edge.'

'It's a long way to the edge of a plane. Infinitely long. The place where parallel lines appear to meet is at infinity. In the Euclidean plane of Flatland, infinity doesn't exist. You can go as far as you like, but you can't actually *get* to infinity. So the horizon *looks* like a line, but it doesn't really exist. When the eye looks at parallel lines, they appear to meet. Parallel lines don't exist in the geometry of the visual system, so we need a new kind of geometry, one in which *any* two lines can meet. And that's what happens on the Projective Plain.'

Vikki found this hard to believe. 'Infinity can't exist. It's just an ideal concept.'

'Really? I think you'd better come with me. We're going lion-tracking.'

•

Twoday 10 Noctember 2099

What the Space Hopper meant, Diary Dear, is that we were going to follow the parallel lions and see where they met. In short, we were going to walk to the horizon! Totally mad, I thought, just as usual: the poor creature doesn't have much of a brain anyway, and what it has seems to have flipped.

Wrong again! I'm beginning to think that the opposite is true: that the Space Hopper's bulbous head is *all* brain, and a first-class one at that. He just *behaves* like a nutcase, for dramatic effect.

Be that as it may, we followed the parallel lions, and something very funny started to happen. (Well, the first thing that happened was that I realized I'd left my right shoe behind. The ointment had been so effective, and my feet had felt so comfortable as a result, that I hadn't noticed.)

I pointed this out to the Space Hopper, but he said not to worry, we'd be back where we started shortly. That seemed unlikely, since we were heading for the horizon, but I've learned not to argue with him.

Anyway, as I was saying, something funny happened. You know that normally, when you walk towards the horizon, it just moves away from you – like a Planiturthian rainbow. But on the Projective Plain the horizon behaves quite differently. It stays where it is, and you get closer to it.

I first realized we were actually *reaching* the horizon when I heard the growling. At first it was so faint that my ears couldn't really focus on it, but slowly the noise grew until it was loud and clear.

I asked the Space Hopper what the noise was. Fool. 'Oh, that's the lions meeting!' it told me. 'Having a conversation.' And you know, when I listened really closely, I began to make out words. As far as I could tell, the lions' conversation went like this:

'Pleased to meetcha!'

'Pleased to meetcha!'

'Pleased to meetcha!'

'Pleased to meetcha!'

And so on. Not a very *interesting* conversation, I told the Space Hopper. 'But a very friendly one,' he said.

Then, suddenly, we straddled the horizon: the Space Hopper was on the far side, and I was on . . . well, you appreciate, Diary, that whatever side I am on is by definition the near side, and that's where I was.

We were there.

I told the Space Hopper that it was a funny edge that could be straddled. I mean, the bit he was standing on *shouldn't have been there.* Well, in his usual smart-alecky manner he insisted on pointing out that if it wasn't there, he wouldn't be able to stand on it . . . I hate that kind of conversation, don't you?

Well, things got a bit heated for a while, but the upshot of the conversation was that on the Projective Plain, the horizon exists, but it's not an edge. In fact, there *isn't* an edge. What there is, is—

•

'Infinityville,' said the Space Hopper, proudly.

There was a notice scratched in the sand. It bore the symbol ∞. Beneath it was written, **POPULATION** ∞.

'What's that figure eight thing?' Vikki asked.

'That's not a figure eight,' said the Space Hopper, 'it's the symbol for "infinity". And that's where we are.'

The sign changed: **POPULATION** $\infty + 1$.

'Ah, a birth,' said the Space Hopper.

The sign flickered for an instant, then reverted to **POPULATION** ∞.

'Uh – a death?' asked Vikki, in a quiet voice.

'What? No, not at all. That would look like **POPULATION** $\infty - 1$. But then it would change back to **POPULATION** ∞ again. You see, infinity *plus* one equals infinity, and infinity *minus* one equals infinity. Infinity, basically, is Where Things Happen That Don't. Which is why parallel lions meet there. The Projective Plain is very similar to Flatland, in a lot of ways, except for this business of parallel lions meeting. Which changes things a lot – for instance, "distance" doesn't make much sense.'

'Is that why we got to infinity so quickly?'

'Of course. Anyway, like I said, on the Projective Plain parallel lions always meet. And, like Flatland, they meet at just one point.'

Vikki digested this. 'Hang on. In Flatland, parallel lines meet at the horizon, which isn't a genuine line. Here, apparently, it is. But even so, in Flatland parallel lines meet at *two* points. One at one end, one at the other.'

'Not on the Projective Plain,' said the Space Hopper.

Vikki turned round. In the *far* distance she could see the two parallel lions converging. She looked at her feet (one still unshod):

the lions met a short distance away. Well, it looked short, but distance has no meaning on the Projective Plain . . . No matter. She pointed out the *other* meeting point for the parallel lions.

'There? No, that's not a second meeting point. That's the same place as here.'

'But it's not here. It's over there.'

'Yes, but over there is actually here. The ends of the lions are one, not two – look, all you have to do is use the zoom facility on your VUE and you'll see what I mean.'

Vikki boosted the magnification and inspected the second meeting point from close up. And what she saw was—

Herself.

Talking to the Space Hopper.

The other Vikki seemed to be inspecting something through the zoom facility of her VUE.

Wild!

'You're saying we're over there as well as over here?'

'No, I'm saying that over there *is* over here. It's the same point. We're just looking at it along two different directions. It's like seeing some object, *and* seeing it in a mirror. One ray of light comes straight from the object; the other ray *bends* when it hits the mirror. Or it's like a mirage, when you see the same object in different directions because hot air refracts light.'

'But you said that these lions are straight.'

'They are. But they still bend. On the Projective Plain, lions can be straight and yet close up. Like a circle – in fact, in a topological sense they *are* circles – except that these circles are *straight*.'

'I don't understand,' said Vikki in bewilderment.

'That's because I'm using Flatland words to describe a projective phenomenon', said the Space Hopper, pompously. 'Let's have another go.' He pondered for a few moments. 'OK. Mathematically, infinity is just an abstract construct, so we can endow it with any properties we want. I happen to want lines to meet at only one point. So I insist that the "two" points at infinity, at either end of a pair of parallels, are considered as the same. It may sound odd, but it works. It's sort of like bending the lines round into a circle, except that they stay straight.'

'Clear as mud!'

'Good. So we can think of the Projective Plain as the usual plane, *plus* a "line" at infinity, *plus* the rule that the opposite ends of pairs of parallels meet the line at infinity in the same point.'

'I just can't visualize it.'

'On the contrary, Vikki, it's how your visual system actually works.'

'Well, I'm having trouble getting it into my head in one piece. And it's not at all clear to me why the Projective Plain has only one side, as you say. The ordinary plane has two sides, top and bottom.'

'Yes, but the top surface and the bottom surface get joined together at infinity because of the rule about the endpoints of parallels being the same', said the Space Hopper. 'Don't try to think about it. Just follow me. We'll keep following the same parallel lions as before, in the same direction, and see where we get to.'

They trudged on. Now Vikki's shoeless vertex started to hurt. She was just about to complain when the Space Hopper called a halt. 'We're there.'

'Where?'

'You tell me. What's that lying on the ground?'

Vikki followed the line of his horns. 'My shoe!'

'So?'

'So we've come back to where we started! You're right, the lions here really do seem to go round in circles!' She glanced back to check: a pair of parallel lions led towards the distant horizon, met, and diverged again. With the zoom facility of her VUE, she could see herself standing next to one of her shoes. Shoes! She seized the accidental castoff and pulled it onto her bare right endpoint—

It wouldn't fit.

She inspected it.

'Hang on, this is a left shoe! The one I lost was a right shoe!'

'Have a look through the VUE,' said the Space Hopper. She did. In the distant image, she was definitely looking at a right shoe.

'What you see in the VUE is the same shoe as the one before your eyes,' said the Space Hopper.

'Then why has it changed into a left shoe?'

'It hasn't. *You* have changed into a left Victoria. Yes, I *know* you're a line, so that ought to make no difference, but you have a sense of left and right, and *that* has changed. Your left shoe has changed into a right shoe, and your right foot has changed into a left foot. It is as though you have passed through a mirror, but there *isn't* a mirror. The Projective Plain is a one-sided surface. We're not standing *on* that surface, as a Spacelander would stand: we are *in* the surface, like drawings inscribed upon it, but going all the way through, like ink on tissue paper. And the geometry of the Projective Plain means that when you slide an object along a lion until it gets back to where it started from, it returns as its own mirror image.'

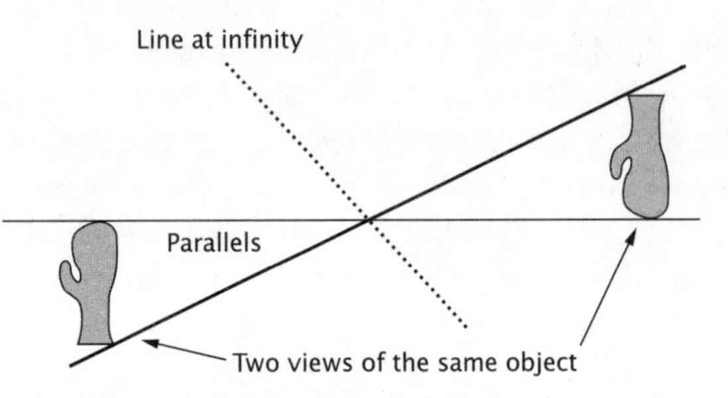

Vikki began to wail. 'I need another shoe! This one won't fit!'

The Space Hopper gave her a kindly look. 'You're tired, too,' he said. 'Give me the shoe, and wait here. I won't be long.' She watched as the Space Hopper headed off along the lions, back towards where they'd come from, carrying her unwanted, wrong-footed shoe. He dwindled into the distance . . .

. . . and tapped her on the shoulder.

She jumped.

'Sorry. I thought you'd see me coming.'

'I was still watching you *going*.'

'On the Projective Plain, coming and going are the same thing. Anyway, I've brought your shoe.'

'The same one?'

'Of course. You don't think I can create shoes out of thin air, do you?'

'Then it's useless. I've tried it and it won't fit.'

'Try it now.' She did. It fitted.

She gave the matter some serious thought. 'Oh! Now I see – if the shoe is carried *once* round a lion, it flips. But carry it round a second time, and it flips again—'

'Getting it back to where it was originally,' said the Space Hopper.

•

Threeday 11 Noctember 2099

Dear Diary,

My experiences with the missing shoe have convinced me that the Projective Plain really does have only one side. It acts like a sort of mirror, but instead of going *through* the looking-glass, you go *along* one.

Moreover, the Projective Plain captures the geometry of *projections* – which are important. For instance, I now see that every time anyone takes a photograph, their camera *projects* the outside world onto the film. The Space Hopper says that by using projective geometry, you can compare a photograph with the real thing and work out where the camera must have been (no, don't ask me how, I gather it's a bit technical). So one-sided spaces are useful after all.

Funny old world, isn't it?

8

GRAPE THEORY

Flatlanders are so *emotional*, the Space Hopper thought. Vikki had gone off in a huff for no reason at all. Just because he was explaining the geometry of the Projective Plain in exquisite detail, that was no justification for being antisocial! Then it belatedly occurred to him that what he thought of as 'exquisite', others might well find 'excruciating'. For once, the bouncy creature felt depressed. He'd offended the little Flatlander by being overenthusiastic. He ought to do something to bring back the sparkle to her endpoints . . .

Of course – the very thing!

He scuttled over to where she was trudging along between the parallel lions. 'I don't know about you,' he said, 'but I could do with a drink. I know a *great* bar just a few paraspatial leaps away in the Mathiverse. Owned by a turtle, has his own vineyard, makes *great* wines – Chordonnay, Modulot, Quadrati, Rhombolo, Bouzo . . . Care to join me?'

Vikki perked up at once. A bar sounded a lot more exciting than a desert. And a paraspatial bar at that! Within a few minutes they were on their way to Running Turtle's, which was situated on a prime piece of surreal estate in one of the more fashionable zones of metaspace. Zone 999999999959, in fact – as prime as they come. Not only that, it had an excellent view of the Galois Fields, the Lakes of Wada, and the Devil's Staircase. They would be able to watch the tangent cones grazing.

•

Elsewhere in the Mathiverse – not that there was a Where to be Else in – the Square family were settling down to their evening meal.

'I do wish Vikki would come home', said Lester, out of the blue.

Jubilee went rigid. It was the first time that Lester had mentioned his sister for weeks.

'What's it like in Numerica, Mum?' asked Berkeley, and Jubilee seized on the distraction like a drowning polygon being thrown a length of rope.

'It's very . . . numerical,' she said. 'They have very big houses and very big cars and they all eat very big meals.'

'Do they have very big cats and dogs?' asked Les, wide-eyed.

'Er – no. I think their cats and dogs are the same size as ours. And they're the usual star shape, too.'

'What sort of big meals do they eat? Giant oxagons?'

'Um . . . giant *lumps* of oxagon. In a bun. They call them hogsburgers, I think.'

'Surely it ought to be oxburgers, dear,' said Grosvenor. 'Hogsburgers ought to be made from hogsagons.'

'I think someone once told me,' said Lee tentatively, 'that hogsburgers are called hogsburgers because they were invented in a town named Hogsburg. Not because they're made from hogsagons. And the Hogsburglars happened to make their hogsburgers from oxagon meat, just to confuse everyone.'

'Well I never,' said Grosvenor. 'Never knew that.' He lapsed into silence.

'Is Vikki eating hogsburgers?'

'Of course,' said Lee. 'Just like any native Numerican.'

'Is she eating them right now, Mum?'

'Probably, dear.'

'And will Vikki be coming home soon, Dad?'

Lee saw the expression on Grosvenor's face, and tried to steer the conversation back to safer lines. 'Not *very* soon, Les, no. Not very soon.'

'Why not? Is she having too much fun?'

Whatever gave the boy that idea? Grosvenor thought.

'I'm sure she's having some fun, Les,' said Lee, 'but definitely not too much! No, she's got to stay there for quite a long time yet, because . . . er . . .'

'Because she's working very, very hard,' said Grosvenor.

'Yes, that's right,' Lee agreed. 'I'm sure that's exactly what Vikki is doing right now.'

•

Running Turtle's was, in fact, run by a large turtle, who poured the drinks and mixed the cocktails himself. It had a pleasant balcony with lots of small, cloth-covered tables, nicely sheltered from the afternoon breezes. The only mildly dissonant note was struck by the large number of small orange snails that were crawling over the patio, the railings, and several of the tables.

'Why is he called *Running* Turtle?' Vikki asked. 'He mostly stands still.'

'Listen.'

Vikki did so, and she heard the turtle muttering under its breath, '. . . makes seven hundred and ninety-six; *five* plus seven hundred and ninety-six makes eight hundred and one; *twelve* plus eight hundred and one makes eight hundred and . . .'

'He's adding things up.'

'Exactly. Keeping a running turtle.' Vikki kicked him.

Running Turtle had undergone an Out-of-Cosmography experience like Vikki – dislodged from his own universe by a random metaspatial fluctuation. He had decided that he preferred the Mathiverse to home, and had set up as a bar-owner in partnership with two other Mathiversian immigrants, Woolly Coati and the Chicken Mock Nugget (who organized the bar food).

It soon became clear that Running Turtle had a problem.

'Blasted Tetrahedro!' said Running Turtle in annoyance.

'Pardon? I don't speak Turtle, I'm afraid—' Vikki began.

'And blast the Enneagono, the Triangulo, the Trapezoidia, the Frustumo, the Chordonnay, and above all the Circumfrovese!' he added heatedly.

'What?'

'Grapes. I've got seven (. . . plus eight hundred and thirteen makes eight hundred and twenty . . .) varieties of grape that I need to test, to see which gives the best wine. I want to plant them in plots in my vineyard overlooking the Galois Fields. Unfortunately my hill is narrow, and I can plant only three (. . . plus eight hundred and twenty makes eight hundred and twenty-three . . .) varieties of grape on each plot of land. And I want to minimize the effects of different soils and different exposure to metaspatial illumination, too.'

'You've got *eight hundred and twenty* varieties of grape?'

'No, just seven (. . . plus eight hundred and twenty-three makes eight hundred and thirty . . .) just . . . just the smaller number I mentioned. I daren't say it, or I'll be forced to add it to the running total. It's a habit of mine, I just can't seem to—'

'In matters of viticulture,' the Space Hopper interjected pompously, 'good experimental design and hypothesis testing are essential to eliminate error.'

'I agree with every word I understand, my friend. And I've drawn up some requirements which I believe will achieve those aims.' With a flourish, Running Turtle produced a sheet of paper on which was written:

> Seven varieties of grape are to be arranged in plots. Each plot contains exactly three different varieties. The following conditions must hold:
>
> - Any two plots have exactly one variety in common.
> - Any two varieties lie in exactly one common plot.

'Very clearly expressed,' said Vikki. 'What's the problem, then?'

'I can't find an arrangement that satisfies my conditions,' said Running Turtle sadly.

'Any ideas?' the Space Hopper asked Vikki.

'None,' she replied, brushing several of the tiny snails to the ground. 'Running Turtle, just be grateful your problem's not about geometry! This crazy Space Hopper will stuff your mind with parallels that meet at infinity, circles that are straight lines, and surfaces that have only one side!'

'Don't you think Running Turtle's problem sounds *geometric*, then?' asked the Space Hopper.

'Geometric? It's about grapes! Which reminds me – I'm hungry.'

There was a *whoosh*, and there appeared a stocky creature covered in feathers and with a rubber beak. It thrust a menu in front of her.

'Oh – thanks. Who are you?' asked Vikki, startled by the rapidity of the response.

'I am the Chicken Mock Nugget. Today's special is sugar-coated rat-on-a-stick followed by a choice of kipper blancmange or pond-weed pudding.'

Vikki made a face, glanced down the menu, and ordered a fruit salad, while privately wondering what she would find on her plate.

'Geometry can be about grapes,' the Space Hopper persisted. 'Mathematics is universal, and its truths apply in many different interpretations. As instance of which: suppose I replace "plot" by "line" and "variety of grape" by "point". Then the Turtle's conditions become:

> Seven points are to be arranged in lines. Each line contains exactly three different points. The following conditions must hold:
>
> - Any two lines have exactly one point in common.
> - Any two points lie in exactly one common line.

See? It's geometry.'

'Hmm,' said Running Turtle. 'But grapes, small and round though they may be, are not points. And plots of land, even if long and narrow, are not lines.'

'True – but irrelevant,' said the Space Hopper. 'We're talking about abstract properties of *arrangements* of objects, not the objects themselves. Logically, it makes no difference what *names* we call them. As the great Planiturthian mathematician David-hilbert said, the logical structure of geometry should make equal sense if the words "point", "line", and "plane" are replaced by "beer mug", "chair", and "table". Names don't matter.'

'I don't think I'd be happy if my beer mug were replaced by a point,' said Running Turtle.

'Look at what *kind* of geometry we've got here. Is it Euclidean?'

'Um – no', said Vikki. 'It's got to be a form of geometry in which there are no parallel lines. Because one of the conditions is that *any* two lines have a point in common. And that condition fails in Euclidean geometry.'

The Space Hopper nodded energetically. 'Right. But that condition holds in *projective* geometry. So Running Turtle's little problem is really about a Projective Plain. A *finite* Projective Plain. It *is* a geometric problem, but the geometry has finitely many points. Or, in this case, baby space hoppers.'

'*Baby space hoppers?*'

'They're the most suitable materials to hand. I could have used grapes, but I don't see any. And for some reason this patio is crawling with baby space hoppers.' He pointed at the things she'd thought were snails. 'Must be the breeding season, though nobody told *me*.' The Space Hopper made a low whistling noise and seven babies – they really *did* look like little shell-less snails, bright orange with tiny antennae – crawled up the leg of the table and arranged themselves in a semicircle in front of him.

'Baby space hoppers!' said Running Turtle in a hopeless voice. 'He's going to solve my wine problem using *baby space hoppers*!'

Numbers from 1 to 7 appeared on the sides of the space hoppers, as if by magic. 'These babies correspond to your seven varieties of grape: Tetrahedro = 1, Enneagono = 2, and so on. OK?'

Running Turtle nodded. He no longer had the energy to disagree.

'Now,' said the Space Hopper, addressing the babies, 'I want you lot to form yourselves into seven lines of three. Got that? OK, then. Right: numbers 1 and 2, just line up somewhere – yes, that'll do fine! Now, another of you – yes, number 6, I saw you volunteer – stand in between them. Good, that's one line of three. Now, number 3, you go and stand somewhere off to the side – that'll do nicely. Number 4, stick yourself in between 2 and 3. And number 5, stick yourself in between 1 and 3. Great, now that's *three* lines.'

'What about me?' squeaked number 7.

'Oh. Uh – just sit right in the middle, will you? Now, if numbers 4, 5, and 6 jiggle about a bit – stay in line with the other guys, OK? – then we ought to be able to get three more lines. Number 4: just shift along a bit so that you're in line with 7 and 1, but stay in line with 2 and 3, yes? Good. Number 5, you do the same kind of thing, in line with 7 and 2 as well as 1 and 3. And Number 6, you get in line with 7 and 3 as well as 1 and 2.

'Excellent! So now we've got six lines. Vikki, what's the seventh?'

'I can't see one.'

'No, I mean: if there was one, what would it have to be?'

'Dunno.'

'Remember Running Turtle's rules? Any two points lie on a line and any two lines meet at a point. Look at number 4. Which babies is he in line with right now?'

'Er – 7, 1, 2, and 3.'

'Not 5 or 6?'

'Well – he's in a line with either of them, of course, but there's nothing else in the same line.'

'So it seems, yes. Suppose numbers 4 and 5 form part of another line. It has to meet line 1–2–6 somewhere, according to Running Turtle's conditions. So where?'

'Well, it can't be number 1 because 1 and 5 already lie on line 1–5–3, and there can't be two lines that contain them. And it can't be number 2 because 2 and 5 already lie on line 2–5–7, and there can't be two lines that contain *them*, either. So that only leaves number 6.'

'Right.'

'Which won't work, because 4, 5, and 6 lie on a circle, not a line.'

'Right.'

Vikki looked pleased.

'And wrong. What did Davidhilbert say?'

'Something about beer mugs . . . oh. There's no reason why a "line" should *look* like a line.'

'No. Especially since it represents a plot of land. We just *deem* (lovely word!) points 4, 5, and 6 to form the seventh "line". And then, by a miracle, we find that Running Turtle's conditions are all satisfied.'

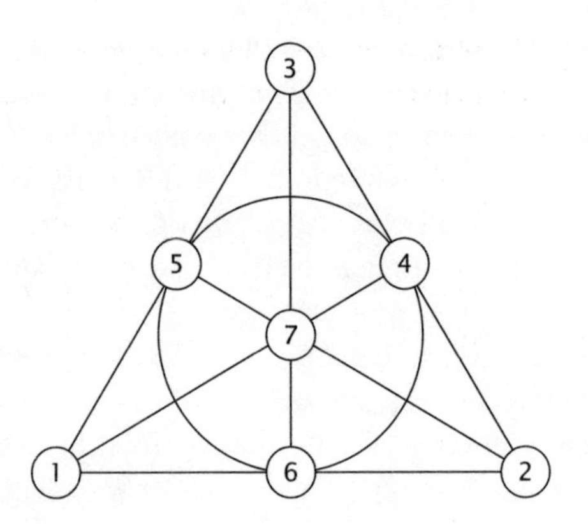

The Chicken Mock Nugget returned with Vikki's food. She was relieved to see that the fruit salad actually was a fruit salad. Raspberry and radish. The chicken sat down next to her.

'In terms of the original problem,' said the Space Hopper, 'I just list the triples of "points" on each "line", and this is what I get:

1–2–6
1–3–5
1–4-7
2–3–4
2–5–7
3–6–7
4–5–6

The numbers 1 to 7 are the varieties of grape. The list corresponds to the seven different plots, each containing three varieties. And both your conditions hold. "Any two plots have exactly one variety in common." Well, the first two plots, 1–2–6 and 1–3–5, have just 1 in common. Plots 2–3–4 and 4–5–6 have just 4 in common, and so on. "Any two varieties lie in exactly one common plot." That's true too. For instance, varieties 1 and 5 lie in 1–3–5, and in none of the other plots. And varieties 3 and 6 lie in 3–6–7 only. You can check them all if you really want to.'

The chicken wasn't quite sure. 'But the *curved* line isn't—'

'It has to be drawn in a curve, Chicken Mock Nugget, because the "points" and "lines" aren't *real* points and lines, they're varieties of grapes and plots of land! Look at the list! It works! Do you *care* that one "line" is bent?'

'Uh – no. It wouldn't stop Running Turtle planting those combinations of grape varieties, would it?'

'Not at all. Finite geometry is all about combinations, you don't actually *need* to draw diagrams.'

'Though the diagram does have its uses,' said Vikki. 'It's a lot easier to remember than a list. And you can *see* that it satisfies all three conditions. You don't really have to check everything so laboriously.'

'Great,' said Running Turtle. 'I'll tell Woolly Coati to get planting tomorrow. In the meantime, perhaps you and your lady-friend would join us in a relaxing wine tasting.'

•

Several bottles later, Woolly Coati happened to mention that he, too, had some grapes to test. 'I've got thirteen varieties, though, and my plots are larger: each can hold four different varieties. But I still want Running Turtle's two rules to hold.'

'Terrific!' muttered the Space Hopper, thinking rapidly. *Thirteen points, in lines of four, each pair of lines meeting in a unique point and each pair of points lying on a unique line. That's another finite Projective Plain!* 'Can do. But,' he added, determined to extract the maximum advantage, 'I'll only tell you if you let me tell it *right*. I want to explain where the answer comes from, as well as what it is.'

'Why?' asked Woolly Coati. 'All I want is the answer.'

'Answers without reasons are magic, not mathematics, Woolly Coati. And knowing where that answer comes from might help you solve similar problems later.' Woolly Coati reluctantly agreed, and everyone settled down for a long afternoon. The Chicken Mock Nugget brought nuts and olives, and the wine flowed . . .

'The starting point is the connection between the usual Euclidean plane and the Projective Plain,' said the Space Hopper. 'To get the Projective Plain you take the Euclidean plane and add an extra "line at infinity", which has one point for each *direction* in the Euclidean plane. If a bunch of parallel lines all point in the same direction, they are deemed to meet at the corresponding point on the line at infinity. Right?'

'I wouldn't dream of contradicting', said Running Turtle, downing a glass of full-bodied red.

'I should hope not! Now, to get a finite analogue of a Projective Plain we need to start with a finite analogue of the Euclidean plane. Vikki, does that ring any bells?'

'This Chordonnay is really, really goo— Oh, sorry, Space Hopper. You . . . said something?'

'Does "finite analogue of the Euclidean plane" ring any bells?' repeated the Space Hopper, a little testily.

'My head is ringing a bit, but no bells, Hopper, not one.'

'The Double-Digit District, Vikki! That was a four-point analogue of the Euclidean plane. Two coordinates – but only taking values 0 and 1, remember?'

'Oh, yes, those e-mail messages.'

'Four points forming a square,' said the Space Hopper, beginning to suspect that no one else was listening. 'Now, we have to decide what "straight lines" should be in such a geometry. That's easy: they're the sides of the square and its diagonals. Finally, we add a line at infinity. The top and bottom of the square are "parallel" in the sense that they don't meet. So we add to each a point at infinity. Similarly for the left and right sides. The diagonals form a third pair of parallel lines.'

'But the diagonals of a square *meet*,' Woolly Coati objected, staring into the far distance with his eyes crossed.

'Not on the grid, they don't,' said the Space Hopper. 'In *this* version of the Euclidean plane, only the four corners of the square exist. The diagonals don't meet at a corner, so the centre of the square doesn't count. Since they don't meet *on the grid*, I consider them to be parallel.'

'Fine by me,' said Woolly Coati, opening another bottle.

'Those three sets of parallels provide three extra points, forming a new line – "at infinity" – and we get a system with seven points and seven lines. Of course it's just a disguised version of the 7-point Projective Plain I talked about before.'

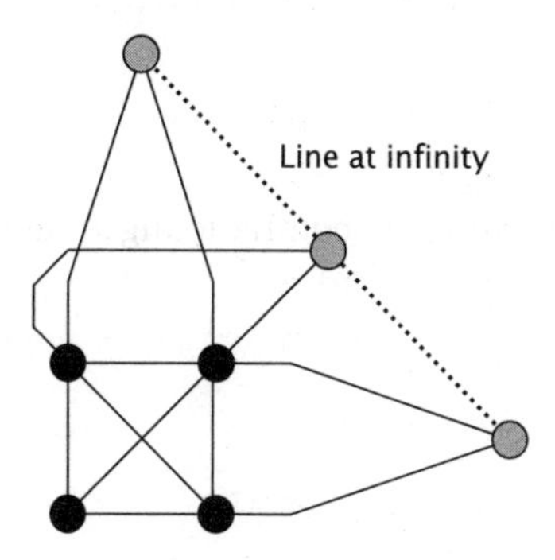

'Wonderful!' yelled Woolly Coati. 'No it's not', he corrected himself. 'My plot has thirteen varieties, so it'll need thirteen points, not seven.'

'Don't worry, I'm coming to that,' said the Space Hopper.

'Soon?'

'*Fairly* soon . . . To get a 13-point Projective Plain we play the same game, but this time starting with a 3 × 3 grid of nine points. The lines in the grid come in sets of three "parallel" lines. The obvious sets of parallels are the horizontal lines 1–2–3, 4–5–6, and 7–8–9 and the vertical lines 1–4–7, 2–5–8, and 3–6–9. But there are two more sets, which "wrap round" – one consists of the broken diagonals 1–5–9, 2–6–7, and 3–4–8, and the other is made up of the broken diagonals 7–5–3, 4–2–9, and 1–8–6. Now all you have to do is add a "line at infinity" with four points, numbered 10–11–12–13, one for each set of parallels. And then . . . you get a 13-point Projective Plain!

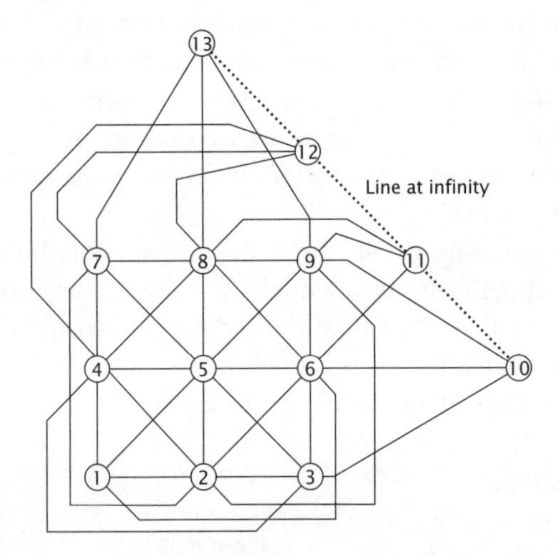

'If you want to, you can read off a list of thirteen plots, each with four varieties of grapes in it, to solve your problem, Woolly Coati.'

'That's amazing!' said the Chicken Mock Nugget. 'Let me try! I'll start with a 4 × 4 grid—'

'Whoops, no, you—'

'—I can do it! Don't interrupt, Hopper, I'll soon get the hang of this . . .'

Two hours later, the sun was beginning to set, and the Chicken Mock Nugget's eyes had a glazed look. The wine had long run out, but that wasn't the reason. 'It doesn't work,' he said finally.

'I either get too many points on lines, or not enough lines through points.'

'I was trying to tell you that hours ago,' said the Space Hopper. 'The method won't work with a 4×4 grid because 4 isn't prime. You have to use a square grid whose size is a *prime* number. You could start with a 5×5 grid, for instance, and you'd get six sets each containing five "parallel" lines. You'd end up with a Projective Plain containing 31 points arranged in lines of six.'

'Where dose numbersh come fum?' asked Vikki.

'Well, the number of points on each line, *not* counting one point at infinity, is called the order of the Projective Plain. And the total number of points, including all those at infinity, is one plus the order plus the square of the order. Each line contains one more point than the order, and the number of lines is the same as the number of points. So if the order is 2, there are three points on each line, and $1 + 2 + 4 = 7$ points altogether. If the order is 3, there are four points on each line, and $1 + 3 + 9 = 13$ points altogether. And so on.'

'Which – *hic!* – ordersh're possible, Hopper?'

'There are Projective Plains of any prime power order – 2, 3, 4, 5, 7, 8, 9, 11, 13, 16, 17, 19, and so on.'

'What about the rest?' asked Woolly Coati. 'Order 6, for instance?'

'The rest are tough. The Planiturthians have known for a century that there isn't a Projective Plain of order 6. You can't arrange 43 points in lines of seven while obeying the conditions that each pair of lines meets in a unique point and each pair of points lies on a unique line. So, Woolly Coati, if you'd had 43 varieties of grape to test, in plots that could hold seven varieties, you'd have been out of luck.'

'No I wouldn't, I'd have been rich.'

'I was speaking hypothetically.'

'So was I!'

The Space Hopper sighed, and battled on. 'There's been one big advance, which rules out orders 14, 21, 22, and infinitely many other values. But that still leaves a lot of orders unsolved – 10, 12, 15, 18, 20, and so on. It's a hard problem because the number of possibilities get very big. When the order is 10, for instance, there are 111 points, which have to be arranged in lines of 11 apiece. There's no way you could try every possible arrangement, life's just too short. However, a People named Lamthielswierczandmackay devised a computer proof that there isn't an order 10 Projective Plain. It took nine years, on and off. Now, here's an attractive challenge for you – the order 12 case is still open! Of course, an attack along the same lines would take ten thousand million times as long, but . . .' He noticed that his audience had gone to sleep.

How could they possibly doze off when he was telling such a gripping tale? Putting it down to the wine, the Space Hopper deflated itself to a comfortable girth, and joined them.

WHAT IS A GEOMETRY?

Fiveday 13 Noctember 2099

How am I feeling?

Somewhat overhung, Diary Dear, since you ask. Which is a remarkable condition for a Flatlander to be in. We usually suffer from stickouters, not hangovers. But I am definitely overhung, not outstuck.

I hadn't realized that finite geometries could have such a drastic physiological effect!

However, that's not why I'm writing this. I had a really important idea somewhere in the middle of the night, and I need to make a note before I forget it. Here is the idea:

WHAT *IS* A GEOMETRY, ANYWAY?

As you will instantly grasp, it is a stroke of sheer *GENIUS!!!* (she said modestly). You see, the Space Hopper has been so busy carting me round dozens of different geometries that until now it never occurred to me that something was missing. Namely, Diary my Pet, an Organizing Principle.

It was the finite projective geometries that did it. Until then, I'd kind of got it in my head that geometry was about spaces, and spaces were sort of continuous. Smeared-out sort of things, right? The kind of environment you could imagine things wandering around in.

Not just a small collection of dots.

Yes, I know the Space Hopper had taken me to the Double-Digit District and on to Triple-Digit Territory, but all that talk of distances along the edges made me think they were just particular bits of Euclidean space, with the emphasis on their corners. Now I see better. Only the relationships between those corners matter: the rest don't have to be present at all. We were looking at finite geometries then, too.

Anyway, some time in the last twelve hours I awoke from a rejuvenating snooze and realized that I now have no idea whatsoever of what a geometry is. So now I'm bugging YOU with it, Dearest Diary, in order to bug the Space Hopper with it once he, too, regains consciousness. Though, to judge by the Doughmouse-like snores emanating from his general vicinity, I wouldn't hold my breath.

•

'You mean you haven't worked it out?' said the Space Hopper in incredulity. 'I show you all these marvellous examples of geometries – plane Euclidean, 3D Euclidean, *n*D Euclidean, fractal, projective, topological, finite projective – to name but a few – and you *still* don't know what you're looking at?'

'That's right. It's just a random grab-bag of unrelated stuff, as far as I can see.'

'Oh dear.' The Space Hopper looked downcast. 'I haven't done my job as well as I thought.' He rubbed his horns together (not easy) in search of inspiration. 'Hmmm . . . ah! Yes! We must temporarily return to the vicinity of Flatland!'

'You mean I can contact Mum and Dad?'

'I really don't think that would be wise at this juncture. It would only upset them, and it might get them into trouble with the authorities.'

'Oh.' Vikki wasn't sure whether to feel disappointed or relieved. Certainly she was having far too much fun to think of going home just yet. But it would have been nice to leave a message. On the other hand, what could she tell them? Anything she said would either have to be a lie, or be totally unbelievable.

A few hours later, subjective time, they hovered a short distance away from the Euclidean plane of Flatland, displaced from it along a metaspatial dimension that can best be captured by the word 'above'.

The Space Hopper had tactfully chosen a region remote from Vikki's house, so as not to risk making her feel homesick. This was somewhere on the edge of a city, and a crowd of polygonal Flatlanders – mostly equilateral triangles and squares, nothing very posh – were making their way through the streets.

'There,' said the Space Hopper. 'Do you see?'

'See what?'

'The essence of geometry.'

'No. All I see is a bunch of polys on their way to work.'

'You don't see a hint of a far-reaching principle that can unify all the different geometries we've visited so far?'

'If it's a hint, it's so subtle that it hasn't registered on my conscious mind, Hopper.'

'But you do see what happens to the polygons when they move?'

'Nothing happens.'

'*Brilliant!*' The Space Hopper bounced with joy. 'You've grasped it immediately! Oh, frabjous day!'

'What in the Plane are you gabbling about?'

The Space Hopper stopped bouncing and its face fell into a ∩. 'You mean – that was just a guess?'

'Hopper, I don't know *what* it was. An expression of ignorance, probably.'

'But you do see that the polygons can move without anything else happening to them?'

'I can't imagine what else *could* happen.'

The Space Hopper's ∪ returned. 'You'll retract that statement in a hurry once I've adjusted the VUEfield to include the part of Flatland we're looking at.'

Flatland ... well, it *shimmered*. Suddenly, the polygons began to distort as they moved. What had previously been an entirely ordinary equilateral triangle *flowed* into a convoluted swirly shape. You couldn't even tell it was triangular. Squares turned into ellipses, Maltese crosses, and irregular hexagons with curved edges. Some shrank, some swelled to ten times their original size.

It was awful.

Vikki tried not to feel sick. 'Hopper! What are you doing to those poor people?'

'Don't worry, Vikki. They don't feel a thing. To them, everything seems normal. It's just the effect of the VUE. Now, tell me what *didn't* happen to them before when they moved.'

'They – they didn't change shape. Or size.'

'Correct. That's what happens to an object in Euclidean geometry when you *transform* it. Effectively you make it move, though technically that's not quite the same thing. Its shape and size are *invariant* under Euclidean motions. That means they don't change.'

The Space Hopper let Flatland slip back out of the VUEfield, and at once everything returned to normal. 'Watch closely. As each polygon moves, it is being *transformed*. Mostly what you're seeing is *translation* – the whole polygon just moves rigidly without any change of its orientation. But what about *that* one?' The Space Hopper pointed to a small pentagon that was spinning endlessly about its centre.

'That one? That's just a kid playing Dizzy.'

'That kind of transformation is called a *rotation*. And there's one more transformation, which Flatlanders can't actually do to themselves. But, with a little bit of metaspatial help . . .' The Hopper leaned down, tucked the tip of one horn under a Hexagon that had stopped near a large public building, and neatly flipped it out of the Plane of Flatland altogether. Screeching in fright, he spun like a tossed pancake, and flopped back into the Plane, landing upside down.

'That's called a *reflection*,' said the Space Hopper. 'If you want to perform a reflection while staying in the Plane, you have to use a mirror. But if a third dimension becomes available you can reflect an object by rotating it in the extra dimension.'

'Oh, poor thing! He'll be most upset!'

'He was. Literally. And I can't leave him that way – all his internal organs will be back to front. All his molecules, too. He won't be able to get any nourishment from ordinary proteins. He'll starve unless I do something. So . . .' The unfortunate experimental subject was flipped out of Flatland a second time, reverting to his normal handedness. He fled down the street, yelling incoherently.

'If you perform a reflection twice, Vikki, you get back to where you started,' the Space Hopper pointed out. 'Isn't that fascinating?'

'Considering what you've just done to that poor Hexagon, I hope there's some *point* to all of this,' said Vikki in a menacing tone. 'He could have been a friend of mine.'

'Point? Oh, yes, very much so . . . What you've been looking at are rigid motions in the plane. Translations, rotations, reflections. Those transformations characterize the geometry of Flatland: the only meaningful geometric properties of Flatland are the ones that are *unchanged* after a rigid motion. Invariant. Properties like length, area, angle, and so on.'

'Oh.'

'Whereas in topology, say, the range of allowable transformations is far greater. Length, area, or angle are *not* topological invariants. They can all be changed by continuous transformations. They can change in projective geometry, too – but in projective geometry the permitted transformations are projections, so straight lines stay straight. "Straightness" is invariant in projective geometry. But it's not invariant in topology, because there you can take a straight line and bend it.'

Vikki mused upon this new idea. 'You're saying that . . . you can distinguish different geometries by their allowable transformations?'

'That's right.'

'But that still doesn't tell me what a geometry *is*.'

'No, but we're nearly there. Let me tell you about *groups*.'

•

Sixday 14 Noctember 2099

Another e-note for you, Diary Dear, before I forget. This mathematical stuff is very cute, but it's a pig to remember.

A space is a set of objects, conventionally known as *points*. But they don't have to LOOK like points, actually. The set of all triangles in the plane, for instance, is a space. Its 'points' are really triangles. It has six dimensions: two to specify the position in the plane of each vertex.

A transformation, I gather, is in effect a recipe for moving things around inside a space. 'Translate by seven units in a northeasterly direction.' 'Reflect in *this* mirror.' And so on.

Transformations can be combined with one another by doing first one, then the other. For instance, think of a square. Rotate it through 90 degrees, then through 180 degrees. That's two consecutive transformations. The result is the same as 'rotate through 270 degrees' – another transformation. According to the Space Hopper, the word for this is that the transformations form a *group* – meaning that the result of performing any two of them in turn is another one.

A geometry is a space together with a group of transformations of that space. For the 2D Euclidean geometry of Flatland, the space is a Plane and the transformations are rigid motions. For topology, the space is whichever topological object you're interested in, and the transformations are continuous deformations. For 2D projective geometry, the space is the Projective Plain, and the transformations are projections. And so on.

The legitimate geometric concepts, for a given space and transformation group, are the things that are *invariant* under all transformations in the group. Things like lengths and angles for Euclidean geometry; things like 'knotted' for topology; things like 'lying in a straight line' for projective.

This idea, apparently, was invented by Felixklein – yes, the Planiturthian who came up with the Klein bottle. Pretty clever guy, eh? What it means is that ALL of the geometries ever imagined turn out to be different versions of the same idea: groups of symmetries of some space. Now *that's* what I call an intellectual synthesis!

•

'How do the *finite* geometries fit into the Felixkleinian picture, though?' Vikki asked the Space Hopper. 'I don't see how you can *move* objects when space is a finite set of dots.' Another objection struck her. 'And what about fractals?'

'Ah. Those geometries are *also* associated with transformations. For fractals, the allowable transformations are lipeomorphisms.'

'Right, that makes perfect – sorry?'

'Lipschitz diffeomorphisms.'

'*OK*, that *explains* it! Well *done*, Hopper!'

'You're being sarcastic,' said the Space Hopper. 'Look, roughly speaking, a lipeomorphism is a transformation that lives somewhere between a topological transformation and a rigid motion.

Unlike a rigid motion, it can change distances, but there's a limit to how far it can stretch things, whereas there's no limit for topological transformations.

'The important thing is that the fractal dimension of a shape is invariant under lipeomorphisms. So a fractal is a geometry, and its fractal dimension is a legitimate concept within that kind of geometry.'

'I believe you. And the finite geometries?'

'There, the transformations are *permutations* of the points. They're ways to change the labels on the points – to rearrange them, if you like – while preserving the relation of "being in the same straight line". For instance, think about the 7-point projective plane. If I rearrange the points so that the labels 1234567 become 4736215, for instance—'

'Where did *that* come from?'

'A few straightforward but messy calculations, Vikki. Educated trial and error, if you like. Anyway, if you permute the labels like that, you find that every line transforms to a line.'

'I don't quite follow.'

'Well, 1–2–6 is a line. The permutation sends 1 to 4, 2 to 7 and 6 to 1. So 1–2–6 goes to 4–7–1, and that's one of the seven lines. You can check all seven if you want.'

Vikki did. He was right. 'Is that the only possible permutation that sends lines to lines?'

'Not at all. There are exactly 168 of them, as it happens.'

'That's a lot!'

'It is. It means that the 7-point projective plane is extremely symmetric – it has 168 symmetries.'

'I'm not sure what you mean by "symmetry",' said Vikki.

'Ah! That's the most important idea of all!' said the Space Hopper. 'What does "symmetric" mean in ordinary language?'

'Ummm . . . well-proportioned? Elegant?'

'Yes. Mostly, people use the word in a metaphorical sense. Sometimes they mean something more specific, though: that an object looks the same when viewed in a mirror. Like your Mother. Mathematicians call this "bilateral symmetry", and see it as the simplest example of a much more general concept.'

'That doesn't surprise me,' Vikki said, sniffing a little. The Space

Hopper, in his bumbling way, didn't realize that mentioning her family might make her feel unhappy. But, actually, it was comforting to talk about them: it reminded her that they still existed. Oh, and so *close* . . .

'Your father, for example, has eight symmetries.'

'That does surprise me. I didn't realize he had *any*.'

'Oh, yes. As long as you ignore fine details like where his eyes are, of course. His general *shape* has eightfold symmetry. Poor Planiturthians only have twofold symmetry – bilateral. And it's somewhat imperfect. Nonetheless, they make an awful lot of fuss about the "beauty of the human form" and suchlike. Well, no doubt it's beautiful to *them* – but to us Space Hoppers you can't beat a good fat bouncy blob with svelte orange skin, a pair of horns, and a cheeky smile.'

'Grin.'

'What's the difference?'

'Total visible area of teeth.'

The Space Hopper quickly changed the subject. 'To mathematicians, a symmetry – note the "a", we're talking specifics here – isn't a thing, it's a *transformation*. To be precise, a symmetry is a transformation that leaves an object looking exactly the same as it was before the transformation was performed. For instance, if someone rotated your dad through a right angle while you're looking the other way, you wouldn't notice any difference in his overall shape. So "rotate through a right angle" is a symmetry transformation of any Square.

If a square is rotated through a right angle . . .

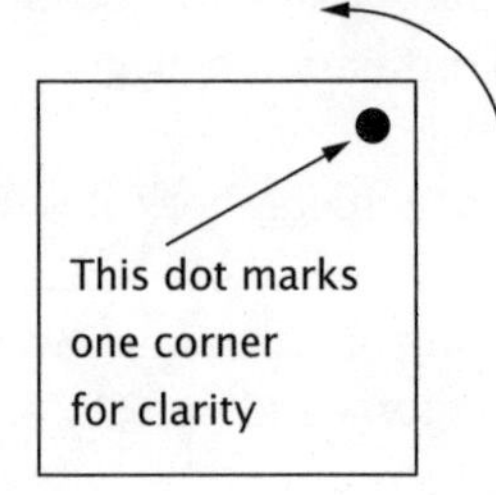

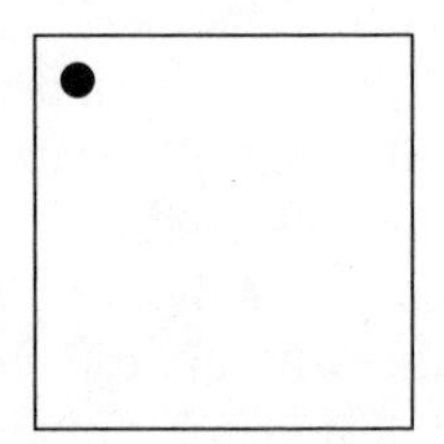

. . . its appearance is unchanged

'Sometimes more than one such transformation exists. Then, the object has symmetr*ies*, not just symmetry. That happens to be the case with your dad, because he's a Square. Every Square in the Euclidean Plane has eight rigid-motion symmetries. Can you think of any?'

'Well . . . let's see . . . "rotate through *two* right angles"?'

'That's right. If "rotate through one right angle" doesn't change the shape, then another "rotate through one right angle" doesn't either, and their combined effect is "rotate through two right angles". Any more?'

'"Rotate through *three* right angles"!' Vikki said, cheering up a little. That was an *easy* one. And the Space Hopper's enthusiasm was infectious.

'Very good. And that's the same as "rotate through one right angle in the opposite direction", of course.'

Vikki stared at him. 'Of course? But that twirls him the other way, and through a different angle.'

The Space Hopper's horns drooped disconsolately. 'Sorry. Silly me. I forgot to explain something extremely important about what a "transformation" is. The only thing that matters when describing a mathematical transformation is where everything ends up, relative to where it started from. The *route* it takes on the way is ignored.

'If your dad rotates through three right angles in one direction – let's call it "clockwise" to be specific, though it would look anticlockwise if you viewed Flatland from the other side – then he ends up in a particular position, and we can tell what that position is by looking at his eyes, even if we can't tell by looking at his outline. If, instead, your dad rotates through one right angle in the anticlockwise direction, then again he ends up in a particular position, and again we can tell what it is by looking at his eyes. Not only that, it's the *same* position as the previous one! So in that sense the transformations specified by "rotate clockwise through three right angles" and "rotate anticlockwise through one right angle" are identical.'

Vikki said she understood. All that matters is how each initial point corresponds to where it ends up. Clear as mud.

'Any other symmetries?'

'Well – "rotate clockwise through *four* right angles"? No, hold it – everything ends up where it started if you do that. So that doesn't count—'

'Actually, it does. "Leave everything where it is" is a transformation, and it certainly can't change the shape! It's known as the *identity transformation* – from "identical", I guess. The identity is a symmetry of any shape whatsoever!' The Space Hopper's grin faded. 'But not a terribly interesting one. And quite often, the only one. When we say that a shape is *asymmetric* – without symmetry – what we really mean is that it is without any symmetry other than the identity. The identity is "trivial", you see. But in a precise theory of symmetry, we have to bear it in mind, trivial or not.'

'Why not rule it out? Define a symmetry to be any non-identity transformation that leaves the shape looking the same.'

The Space Hopper pursed his rubbery lips. 'Could do. But then you destroy one of the key properties of the symmetries of an object, the "group property": any two symmetry transformations carried out in succession lead to a symmetry operation. If the first transformation leaves the shape unchanged, and the second transformation leaves the shape unchanged, then doing them both in turn *must* also leave the shape unchanged. Yes? If you don't change the shape, and then you don't change it *again*, then you haven't changed it!'

'That's kind of obvious, isn't it?' asked Vikki.

'Sure. Obvious things can still be true – though often what *seems* obvious turns out to be false, young Victoria. The fact that two symmetries make a symmetry is both obvious and profound. It leads to a kind of "calculus of symmetry", known in the Mathiverse as group theory. But I'm getting ahead of myself: you've still only told me four of your dad's symmetries, and I reckon he has eight.'

'Can't see any others. I mean, "rotate through five right angles" is the same transformation as "rotate through one right angle", if I've understood what you've just told me.'

'Absolutely.'

'Then I'm stuck.'

'I think you may find that, on reflection, you're not.'

'But I am! I just told you so! More thinking isn't going to help—'

'That's not what I said. I said that I thought that *on reflection* you—'

'Oh! That was a clue, wasn't it? Reflection! If Dad stands in front of a mirror, his reflection still looks like a Square!'

'Yes – although with most mirrors, his position in the Plane changes. However, there are four mirrors which, after reflection, leave his outline exactly the same as before.'

'I don't see how that can be true. If he stands in front of a mirror, then his image will appear to be behind it. Wherever the mirror is. And if he stands behind it, there won't be any image at all.'

'Yes, but these are mathematical mirrors, not glass ones. You don't have to stand in front of them for them to reflect you. You could stand behind them and they'd still work.'

'That doesn't help either. If Dad stands behind a mathematical mirror, then his image will appear to be in front of it.'

'Right again. Now, think like a mathematician. Are "stand in front of" and "stand behind" the only possibilities?'

'Yes! Wait, wait, no! You could stand part-way across the mirror – some parts of you in front, others behind.'

'Yes. And if you're Grosvenor Square, and you want your mirror image to coincide with you, just where do you stand relative to the mirror?'

'Well – the mirror has to run through your middle,' said Vikki, 'or else the image pokes out differently from you, if you see what I mean.'

If a square is reflected in a mirror,
then it only fits its original outline . . .

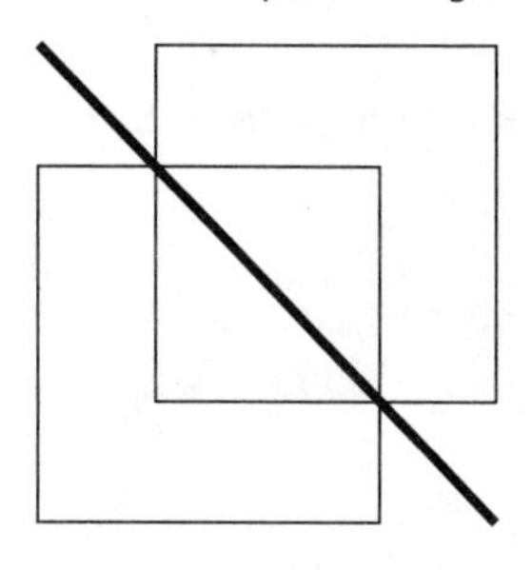

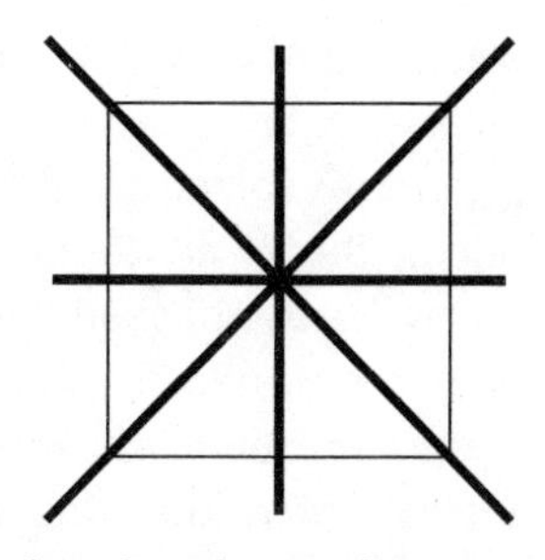

. . . for these four choices of mirror

'Very good. Will *any* mirror through Grosvenor's middle give an image that coincides with him?'

Vikki called up the VUE's menu-bar visualization routine to try a few thought experiments. 'Oh, no. The image can flop over into a different position, even when the mirror runs through the middle of the Square.'

'Right again. But are there any positions for the mirror where that doesn't happen? I'll give you a clue: what happens to the angle between one corner of the Square and the mirror, when the Square gets reflected?'

Vikki doodled with the VUE again. 'It *doubles*.'

'Correct. But if the Square's image fits on top of the Square, what can that doubled angle be?'

'Oh! A right angle!'

'Yes. Or two right angles, or three right angles – or no right angles, zero degrees.'

'So twice the angle between a corner and the mirror has to be a multiple of a right angle . . .' Vikki mused. 'And the angle itself must be a multiple of *half* a right angle – which is 45 degrees.'

'Beautifully expressed – I couldn't have put it better myself! So the possible angles between the mirror and a corner of the Square are 0, 45, 90, and 135 degrees. Where do those place the mirror?'

'Um. Zero degrees puts it . . . along a diagonal of the square. Then 45 degrees is . . . through the middle, parallel to two of the sides. Ah! And 90 degrees is the *other* diagonal of the square; and 135 degrees is through the middle, parallel to the other two sides.'

The Space Hopper waggled his horns in delight. 'So you've found – how many reflectional symmetries?'

'Four.'

'Which together with three nontrivial rotations and the identity makes . . .?'

'Eight. Like you said.'

'Good!' The Space Hopper bounced energetically up and down. 'So now we've established that Grosvenor has eight symmetries—'

Vikki interrupted. 'Hang on, you're hopping to conclusions.'

'I am?' The bouncing ceased.

'You told me to think like a mathematician. How do we know it's *only* eight?'

'Choose a corner. You can place it on top of any of the four corners, itself included. Having done that, there are only two ways to position the neighbouring corners: rotate the square, or reflect it about the diagonal through your chosen corner. That makes eight ways to reposition it.'

'It's obvious now you've told me. I withdraw my objection.'

•

Sevenday 15 Noctember 2099

Dear Diary,

I have just had a BRILLIANT idea!!! I now know *why* our ancestors placed so much weight on how many sides a polygon had, and how regular it was.

It's not really about the number of sides *as such*, and it's not regularity *as such*.

It's about SYMMETRY.

For the same reason that Dad has eight symmetries, every Pentagon has ten, every Hexagon twelve – the number of symmetries of a regular polygon is twice the number of its sides. A genuine Circle has *infinitely many* symmetries – no wonder the Priests were so snobbish! Of course, an approximate Circle, which is what they really were, has perhaps a hundred sides . . . but even so, that's *two hundred* symmetries, which is a lot.

On the other hand, a poor little Equilateral Triangle has only six symmetries – three rotations, three reflections – and an Isosceles Triangle has only two: the identity and a reflection. An Irregular Triangle of the 'criminal classes' has *only the identity symmetry.*

An interesting point for future reference, Diary Dear: Flatland women have *two* symmetries: the identity and rotation through 180 degrees. In fact – a technical caveat that may or may not be significant – there are *four* rigid motions in the Plane that leave a woman's form invariant. Two of them I've just described. The third is to *reflect* the Plane in a mirror that runs along the woman's length. And the fourth? Reflect in that mirror *and* flip her end over end. (I mean rotate her by 180 degrees. Didn't mean to sound risqué!)

So, with the exception of women – misunderstood, as always, by the male-oriented society of a century ago (and it hasn't changed MUCH, Diary Darling, as you and I both know) – anyway, with that exception, one's social position was determined almost entirely by *the size of one's symmetry group*!

This, of course, must be the result of sexual selection. The Space Hopper says the same thing happens on Planiturth – female birds prefer males with symmetric tails, for instance. As I understand the process, a symmetric form is an external sign of 'good genes' – an indicator of genetic fitness. Females that happen to mate with symmetric males (and vice versa, natch) have fitter offspring, so not only are genes for symmetry passed on to the kids, *so are genes for preferring symmetric mates.*

Astonishing. Suddenly the whole of our past snobbish society makes complete evolutionary sense. Bummer!

Except for the lowly role of women, who – as I've told you – are about as symmetric as a Square. Unfortunately, evolution treats males and females differently.

Until now.

We have nothing to lose but our genes!

10

PLATTERLAND

Vikki had the good sense to keep this particular insight to herself. She didn't want to get involved in gender discussions – she was having enough trouble coping with the challenges posed by the Mathiverse. And now the Space Hopper was determined to carry them away to pastures new.

Once more Vikki experienced the indescribable but unforgettable sensation of Space Travel. The Space Hopper grunted in satisfaction. 'Now *this* is a space I really like. Significant. This one will surprise you, and no mistake.'

So far every new space had surprised Vikki, so she couldn't see why the Space Hopper was more fascinated by this particular world than by the others they'd visited. But she had to admit that it was rather elegant. It floated directly below them, a perfect circle, glowing salmon-pink and lavender against the non-coloured backdrop of metaspace. It looked like a giant dinner-plate.

'Welcome to Hyperbolica,' announced the Space Hopper. 'You'll feel at home here: it's two-dimensional. Colloquially known as Platterland to us Space Travellers,' it added, confirming her mental image.

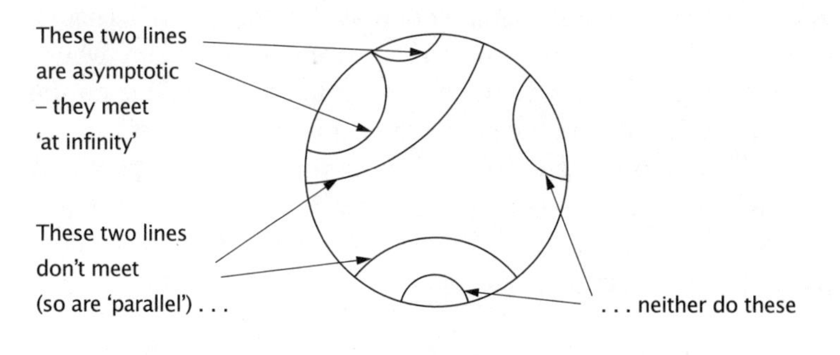

As they descended towards the glowing disc, Vikki began to make out details. The growing expanse of the plate was criss-crossed with curved lines, which at first she took to be roads, until a series of smoke puffs drew her attention to the long vehicles that were making their way across the Platterland terrain.

The curves were railway lines. But there was something terribly wrong with them.

'Hopper, am I right in thinking that those lines are railway tracks?'

'Yes. From a distance they look like single lines, but up close like this they come in pairs.'

Unlike Spaceland trains, the ones in Platterland were running *between* the rails. So was the trail of smoke they left behind. But that made good sense to a Flatlander: in two dimensions, there wasn't an 'on top' direction. No doubt sections of rails, the wheels, and parts of the carriage walls could be dismantled when the train reached its destination, so that passengers could get in or out. That's how Flatlanders would handle such technological obstacles. The smoke could be let out the same way. Or maybe it would just fade away of its own accord . . . But one thing bothered her.

'If those are railway lines,' she said, 'then why are they curved? Surely they ought to be straight.'

'Trains can follow curved tracks, Vikki.'

'I know. But curved tracks hundreds of kilometres long must be terribly inefficient. Why aren't they following the shortest paths?'

The Space Hopper considered this carefully. 'I think they *are* following the shortest paths. The tracks look perfectly straight to me.' He was about to qualify this remark when Vikki objected.

'Straight? *Straight?* Rubbish! They're curved. All of them! They look like arcs of circles. Shouldn't railway lines be parallel? That is, unless the people here know how to manufacture trains whose wheels can get closer together or further apart, depending on the width of the rails. Here the pairs of rails meet at the edge!'

'As I was about to add when you interrupted, it all depends on what you mean by "straight" and "parallel",' said the Space Hopper, 'and "line", for that matter. Each railway track here has one straight rail and one rail that's equidistant from it but not straight.'

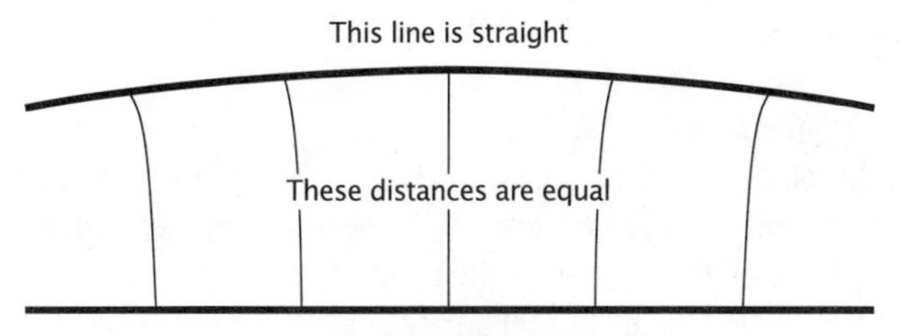

'That's ridiculous.'

'No, not really. Since the equidistant lines aren't straight, they're not true parallels. On the other hand, some straight lines here *are* parallel – but not equidistant. In fact, technically speaking, the distance between parallel lines gets closer and closer to zero.'

'You mean they meet?'

The Space Hopper was baffled. 'Sorry? *Meet?* Where?'

'At the edge.'

'What edge? There isn't an edge!'

'The edge of the plate.'

Now the Space Hopper understood the source of her confusion. 'Ah. Yes, it does look like an edge, doesn't it? But Platterland has no edge. Or at least, if it does, none of the inhabitants can ever reach it. And appearances are deceptive. Some of the lines that look curved are actually straight, and the curves that seem to meet at the edge are really parallel straight lines. But, in this space, the distance between parallel lines is *not* constant. So if you drew a line at a fixed distance from a straight line, which is what you need to make a railway track, then the second line wouldn't actually be straight!' He saw Vikki's look of puzzlement. 'Mind you, if the distance is small, they look pretty much like straight lines do here. Which is to say – from *our* point of VUE – curved.'

'But the lines you're telling me are equidistant get closer together as they head towards the edge, Hopper.'

'Do they?' asked the Space Hopper in a quiet voice.

'Yes, they do!'

'You realize that if you're right, then the trains must shrink as they get closer to the edge? And they can't just get narrower, either, because if they did, they'd stop being train-shaped, and the natives would notice. So the whole *train* must shrink as it approaches the edge. The passengers, too.'

Vikki looked down at the Platterland rail network from her overhead VUEpoint. 'They *do.* Shrink, I mean. They get smaller and smaller the closer they get to the rim.'

'Really?' The Space Hopper's scepticism was palpable.

'I can see it with my own eyes,' Vikki protested.

The Space Hopper sighed. 'Haven't I told you over and over again not to believe what you see when you look at a space from the outside? As external observers we have a privileged position, which is denied to the inhabitants. To them, what they can perceive is what's real – and *all* that's real.' By now Vikki was looking a bit sulky, and the Space Hopper bounced apologetically a few times, his horns wobbling. 'Come with me *into* Platterland, and All Will Become Clear. I promise.'

Vikki's stomach lurched as the bottom dropped out of her universe. *Here we go again!* Platterland rushed up to meet them . . . There was a sound like supersonic suet pudding encountering a mountain range – then the world fell to pieces in total confusion.

•

When Vikki's senses returned, everything seemed perfectly normal. Just as in Flatland, the Space Hopper had reverted to being a mere two-dimensional section of its Spaceland glory. At the moment it was roughly circular, though it shrank and expanded in an extremely disturbing, though rhythmic way. She realized that must be the Space Hopper's breathing.

Actually, things weren't *quite* normal. She felt . . . well, more *curvy* than usual, as if her endpoints had been sucked in. But she didn't feel any unusual compression of her internal organs. It was almost as if her central region had *bent.*

Platterland no longer looked like a plate. No matter how hard she tried, she couldn't see the edge any more. Everything just faded out into a distant haze.

They were standing on a footpath, not far from a level crossing. From out of the hazy distance came the mournful hooting of an approaching train.

They hurried over to the crossing. As they neared it, the barrier slid across to bar the way. Fortunately it was semitransparent, otherwise it would have blocked their view. The rails were semi-transparent too – and the barrier was actually a section of rail, which could be swung aside to let travellers pass when there were no trains coming.

With a sound like thunder the train surged past. When the train was far enough away for it to be safe to cross, the barrier opened. The Space Hopper led Vikki out into the middle of the track, and they stared at the departing train. The smoke had cleared, and she could see it perfectly well, even though the rails had appeared to curve when she had seen them from outside Platterland.

The Space Hopper 'rose' to a level where his eyes could see the train too. 'You're right,' he said. 'Look, the train's shrinking!'

Vikki laughed. 'Don't be silly. It's all perfectly normal. That's just perspective! The further away the train gets, the smaller it looks. It hasn't changed size in the least.'

'Oh?' said the Space Hopper. 'Are you sure?'

'Of course I am!'

'That's not what you said just now when we were privileged observers.'

'But then I could *see* – oh.' This was going to be embarrassing, she could tell.

'Exactly. And you can *still* see, but now you don't take what you see literally. So why did you take it literally before?'

Vikki blushed pale grey. 'Force of habit. But you do have to admit that it all seems perfectly normal now that we've entered Platterland.'

The Space Hopper gave her a mischievous look. '*Me? Have* to admit? Vikki, I am the Space Hopper – *I* don't have to admit *anything*!' He paused. 'But maybe you're right. Let's carry on along this path and see if we can find anything to change your mind. Or mine.'

The path led through a wood, with what looked like perfectly normal flatbushes and jomma trees. An edgehog scurried across the trail just ahead of them, and squarrels ran round the bases of the trees looking for nuts. At least, they *looked* like squarrels. Except there was something funny about them.

The Space Hopper began to shrink. As the circular cross-section of his fat body gave way to the twin circles of his horns, he sank below the plane of Platterland. Vikki just caught a faint cry of 'I'll be back momentarily' before the words floated away on the breeze. There was a flurry of activity and a muffled squeak. Moments later, the Space Hopper was back. Trapped between his horns was one very surprised squarrel.

The Space Hopper was triumphant. 'They're almost impossible to catch, these things. Unless you sneak up on them along an unexpected dimension. I used the nineteenth, did you notice?'

'Don't be silly. Anyway, what's so special about trapping a squarrel?'

'You tell me. Do you *see* anything special?'

Vikki gave the wretched animal a quick once-over. 'It's a perfectly ordinary squarrel, just like the ones in Flatland. It's small and furry, it's square, and it hoards nuts.'

The Space Hopper bobbed in excitement, so that he looked as if he was breathing very rapidly. 'Are you *sure* it's square?'

'Well, admittedly I can't see all its sides without walking round it, but I can recognize right angles when I see them. And when it was running round the tree, all *this* squarrel's angles looked right to me. And all its sides looked the same length, too.'

'I agree. They are. But did you count its sides?'

Vikki exploded. 'Now you really are being silly. If all the corners are right angles and all the sides are equal, it's *got* to be a square.'

The Space Hopper grinned. 'Sure?'

'Absolutely!'

'OK. Just indulge me. Keep an eye on the animal while I pirouette, and count how many sides it has. You can tell you've got back to the start when its nose comes round again.'

Vikki stamped her vertices in frustration – this was *such* a pointless game. 'Oh, if you insist. What a stupid idea! OK, start turning . . . One, two, three, four . . . five. See, it's a perfectly ordinary sq— *five*? Hang on, you're cheating. Turn round again. One, two, three, four, five. Again: one, two, three, four, five. Nuts. How are you doing that? This is mad! How can there be a five-sided square?'

'Maybe it's a right-angled pentagon,' the Space Hopper suggested, with a wicked ∪.

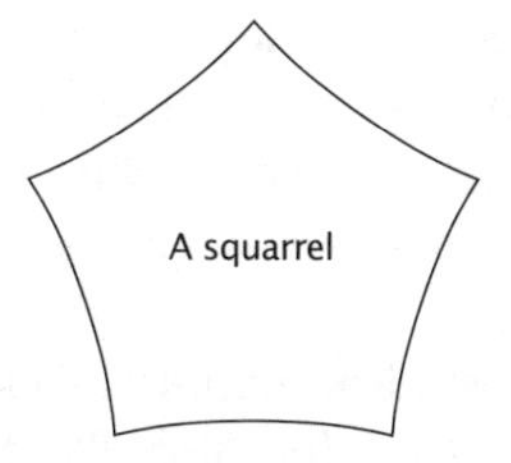

'Yeah, maybe – oh, you're awful. That doesn't help at all, it's just as bad. This is crazy.'

The Space Hopper bobbed in agreement, then stopped. 'On the other hand, Vikki, it may just be showing you that Platterland isn't quite as perfectly normal as it appears. After all, it *did* look a bit weird from the vantage point of Grand Unified Metaspace.'

'But you just said I shouldn't trust my eyes.'

'True. You really must learn not to take everything I tell you so seriously. But not right at this moment.'

•

Wunday 16 Noctember 2099

Diary Dear, you must believe me when I tell you that this Platterland space is as nutty as a squarerootcake. At first everything looks normal, and then when you take a more careful look it's all completely MAD. There are triangles here whose angles don't add up to 180 degrees, for instance. In fact they're ALL like that, though with the smaller ones you'd never notice. The area of a Hyperbolican triangle is proportional to the difference between the angle-sum and 180 degrees: if the sum is close to 180, then the area is close to zero! The bigger the triangle, the weirder it is: there are some triangles whose angles add up to *zero* degrees. Though apparently those aren't really triangles, just shapes made from three straight lines, because their vertices are on the edge, which doesn't exist . . .

I *think* that what saves my sanity is the belief that what counts as a straight line in Platterland is actually curved . . . but the Space Hopper insists that this is merely a parochial Flatland view, or more accurately a parochial *Euclidean* view.

•

'Think of it this way,' the Space Hopper was saying for the dozenth time. 'Imagine that Platterland really is a circle in Flatland, as your visual sense wants you to believe. What kind of fancy physics would make the Platterland railway lines and creatures behave the way they clearly do?'

'I haven't the foggiest idea,' admitted Vikki.

'Fog? Oh dear me, no, not fog. More like a touch of *frost*. Here's a clue: to you, everything in Platterland seems to *shrink* as it approaches the rim of the disc. What physical influence makes things shrink?'

Vikki thought about this. *Pushing* makes things shrink if they're elastic, but somehow she doubted that was the physical influence the Space Hopper had in mind. *Frost?* Frost is cold . . . Ooooh. Things shrink when they get cold, just as they expand when they get hot.

'Cold.'

The Space Hopper's thick orange skin wobbled in encouragement, like hyperactive blubber. 'Very good. Now, as I say, this is only an analogy. The inhabitants of Platterland don't actually feel any kind of cold. But suppose there is some kind of "temperature" effect of which they are unaware: the rim is very cold indeed, and the centre is a lot warmer. Imagine that the temperature falls away according to the distance from the centre—'

'How, exactly?'

'It doesn't matter much, but if you must know, the temperature should be the difference between the square of Platterland's apparent radius and the square of the apparent distance from the centre . . . anyway, assume that all objects shrink in proportion to this "temperature". At the rim, the temperature is zero, so objects shrink to zero size as they approach the rim. Got that?'

'Yes.'

'Fine. So here's my first question. How far is it from the centre to the rim, *as experienced by a Platterlander*?'

'It's just the radius of the disc . . . oh, no, because to them everything shrinks in accordance with the temperature. So let's suppose there is a Platterland creature that can take very big strides—'

'There is. It's called a bigstrider.'

'—and to keep the argument simple, let's imagine that when it starts from the centre its first pace takes it halfway to the rim, as it would appear to us.'

The Space Hopper grinned manically. 'Bigstriders aren't *that* big, but you're on the right lines. Keep going!'

'OK. Um . . . Because distances shrink as the temperature falls, its *next* stride takes it halfway from where it is to the rim – that is, three-quarters of the radius as we would see it. And the third stride also takes it halfway from where it is to the rim . . . seven-eighths of the radius. With the fourth stride, it gets to fifteen-sixteenths of the radius; the fifth takes it to thirty-one thirty-seconds, and so on.'

'Well . . .' said the Space Hopper, with a worried look. 'Your basic idea's right, but your numbers aren't. You're assuming that the temperature is proportional to the distance to the rim, whereas the correct temperature rule would give something distinctly more complicated. Never mind, though: qualitatively, you're smack on target. The length of the strides shrinks *very* rapidly as the bigstrider nears the rim. With your numbers, how many strides would it take to reach the rim?'

'Um . . . *Oh!* It never gets there!'

'Correct. And the same goes for the accurate numbers. Mind you, to our eyes it will appear to get *extremely* close if it keeps going . . . but the poor animal is taking tinier and tinier steps – again from our prejudiced viewpoint – so to *it*, the rim seems just as far away as ever.'

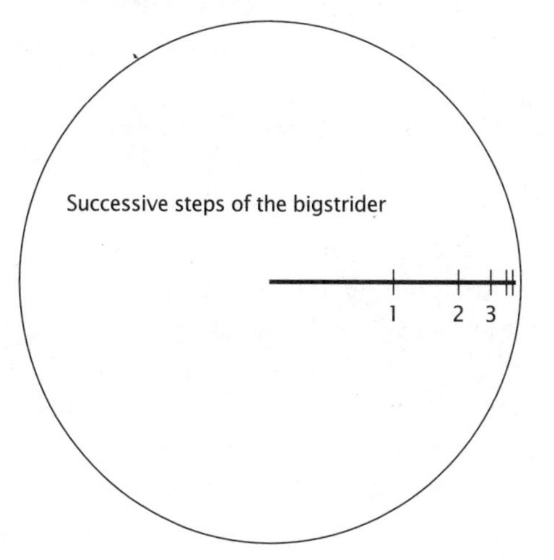

Vikki had a sudden moment of insight.

'The rim is *infinitely* far away, then.'

'From the bigstrider's point of view, yes. And it's *his* world, so we ought to respect his viewpoint.'

Vikki thought about this. 'You keep saying that being outside a space gives us a privileged – and therefore potentially misleading – position. *We* perceive Platterland as a finite disc . . . but its own inhabitants perceive it as being of infinite extent.'

'Exactly.'

'But it's *really* finite, isn't it?'

If there had been any sand around, the Space Hopper would have gone and buried his head in it. And he was *all* head. But there wasn't any sand, so instead he just said, 'Let's go and ask them.'

•

They had landed in a rural part of Platterland, and it took a while before they encountered a native Platterlander. He had that same disturbing appearance as the squarrel. His angles were those of a hexagon, 120 degrees – but he had seven sides.

The Space Hopper introduced himself with a degree of politeness that Vikki had never heard from him before. 'Very formal, Platterlanders,' the Space Hopper whispered to her. 'They expect courtesy.' Then it turned once more to the hexangular heptagon. 'Please excuse my brief conversation with my fair companion, good sir,' he said. 'I was enlightening her with regard to the customs of your delightful country.'

'Ooh aar,' replied the heptagon. 'Tha's a real gennelmun an' no mistake. An' a stranger to these parts, ooh aar, that ye are.'

'That is most exceedingly perspicacious of you, sir. We are travellers from a distant land, visiting your beautiful and prosperous country for the first time. Such is our ignorance that we have many questions – but perhaps it would not be polite to ask.'

'Foire away,' said the heptagon. 'Oi'm bein' a fair bit more broadminded than most.' He paused. 'That's what comes from 'avin' a bigger *'ead* than most, har-har-har!' Clearly this was a favourite joke, and equally clearly it was not to be taken literally – though with seven sides, the heptagon could hardly fail to have a big head.

'I'm curious about your weather,' Vikki began. 'Do you find it cold out here near the rim?'

'Cold? Why no, m'darlin', it's a glorious summer's day. But what's this faffle about a rim?'

'The edge of your world—' Vikki began, and stopped when the Space Hopper kicked her. He would have kicked her in the angle, but she didn't have one, being a Line. Not until now, anyway.

'Edge? You'm thinkin' of a *'awthorn* 'edge, p'r'aps? Not 'ere-abouts, Oi runs a *modern* farm mysen, an—'

Vikki couldn't restrain herself any longer. 'No, "edge" as in "boundary". Where the world ends.'

The heptagon broke into peals of laughter. 'The world 'as no *edge*, m'dear, as you can plainly see for yoursen, it bein' such a glorious day an' the pollution 'avin' been cleared by last noight's thundershower.'

Vikki ignored the Space Hopper's increasingly urgent kicks. 'But if you keep walking away from the centre, don't you eventually reach the edge of the world?'

'Har-har-har, funny joke. But what d'ye mean by *centre*, duckie?'

'The midpoint of the world,' said Vikki, 'the special place where everything is at its biggest.'

'Ah, you'm be meanin' Boondock, no doubt. That's the town where we sells our produce, 'bout two hours by schmule, that-aways. They got buildin's with more'n *foive* rooms in Boondock – that's what Oi calls *big*!'

'No, I mean the place that's different from any other, where objects reach their greatest size!'

The heptagon wrinkled one edge in perplexity. 'No such place,' he finally stated. 'Apart from the local lan'scape, every place on Oiperbolica is 'xactly the same as any other. And everythin' stays the *same* soize – wherever it moight be!'

'But—' Vikki began.

'I think my young companion is referring to the effects of perspective,' said the Space Hopper, diplomatically. 'Things look smaller when they're further away from you', it explained when the heptagon's edge began to cloud over in puzzlement.

'No,' Vikki protested, 'I mean that things really *are* smaller—'

'Nonsense!' said the heptagon. 'If fings changed soize as they moved aroun', nothin'd fit prop'ly any more.'

There seemed to be no answer to that.

•

They had found an inn, The Sober Newt, and were arguing Platterland geometry over glasses of the local firewater. Grosvenor would have been horrified had he known, just as he would have been when Vikki had taken part in Running Turtle's wine tasting, but Daddy was a world away. If not several.

'I don't see the problem,' said the Space Hopper. 'After all, in Flatland, if you watch an object moving away from you, your eyes tell you it's shrinking. In fact, as it heads off into the infinite distance its apparent size shrinks to zero. So Flatland objects behave just like Platterland ones!'

Vikki wasn't having any of this nonsense. 'But on Flatland we know that really the object *isn't* shrinking. It's just the effect of perspective.'

The Space Hopper wasn't buying that. 'Hah! So when your eyes tell you that *Flatland* objects are shrinking as they move off to infinity, that's just an optical illusion. But when your eyes tell you that *Platterland* objects are shrinking as they move off to infinity, that's reality. Seems to me you're just choosing whichever interpretation suits you best. That's poor logic.'

'But Flatland objects *don't* shrink! If you walk away from somebody else, it may appear *to them* as if you're shrinking . . . but you don't feel yourself getting smaller. You're still the same size relative to your surroundings.'

'But the person watching you sees the surroundings shrinking too. So *of course* you wouldn't notice that you were shrinking.'

A look of horror passed over Vikki's edge. This was starting to sound quite convincing, if you took it at face value. 'You're just teasing me, aren't you?'

'No,' said the Space Hopper. 'I'm trying to put you into a Platterlander's frame of mind about Platterland.'

'I still think that's different. Look – on Flatland you can carry a ruler with you and measure your size. It stays the same, however far away you move.'

'That's because the ruler shrinks at exactly the same rate you do.'

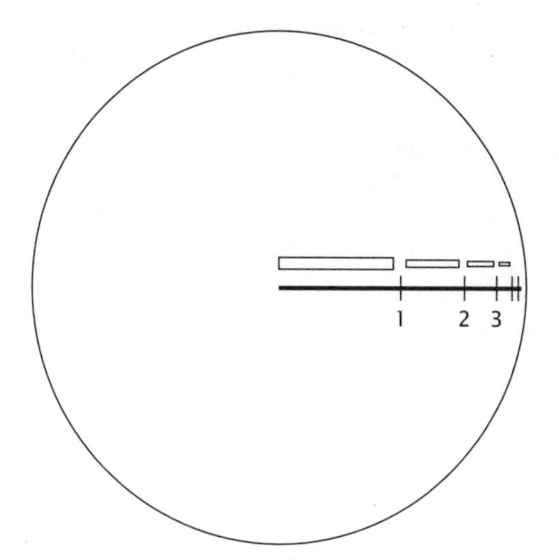

'Rulers can't shrink! Rulers are rigid.'

'So what happens if a Platterlander carries a rigid ruler with them as they head out towards the rim? Won't they notice that their size, as measured by the ruler, doesn't change?'

'But their rulers shrink! The temperature makes *everything* shrink.'

'Just as the temperature in Flatland makes everything shrink as it moves towards infinity?'

'Don't be silly. There's no temperature in Flatland – well, not like the one in Platterland.'

'How do you know?'

'We've never noticed one.'

'But the Platterlanders don't notice *theirs*, either. In fact, it's not real. It's just an aid for Flatland thinkers to understand Platterland geometry.'

Vikki's mind started to rearrange itself. Just like Topologica, Platterland was finally beginning to make its own kind of twisted sense. You just need twisted logic to appreciate the twisted geometry. It might *sound* crazy, but everything was perfectly self-consistent. '*Ooohhhh* . . . I *see*. You're saying that if rigid rulers can change their size as they move – whatever that means – then you can't rely on the rulers to measure distances accurately.'

'Nearly. Actually, I'm saying the exact opposite. Since the only way you can measure distances is by transporting supposedly rigid rulers around, you have to believe what they seem to be telling you. And on Platterland, what they tell you is that nothing *really* shrinks when it moves . . . and the world is indeed infinite in extent.'

Vikki saw the logic, but she still couldn't shake off the feeling that her own perceptions were the true reality, and said so.

The Space Hopper thought about that for a while. 'The trouble with perceptions,' he said, 'is that they're processed by brains. And brains have evolved short cuts in processing as an aid to survival. Often what we perceive is *different* from reality. If you switch your VUE to Spaceland mode, I can make that very clear to you.' He waited for her to do so.

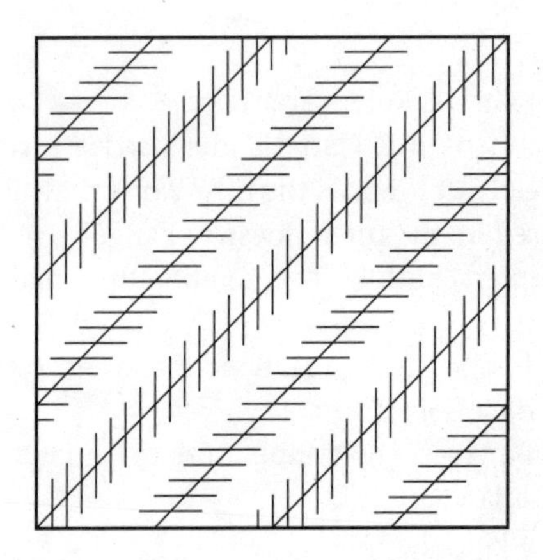

'Look at these lines – the long ones, not the short cross-hatchings. Tell me, do you perceive them as being parallel, or as converging?'

'Obviously they're converging.'

The Space Hopper agreed that it did look like that. 'Measure the distance between them. What does that tell you?'

'Um . . . Wow! It says they're really parallel! But they *look* . . . oh, I get your point now. They're not what they look like. The cross-hatchings fool the eye.'

'They fool the brain into misreading what the eye tells it. I'm not saying that your perceptions and senses are *false*, Vikki, just that there are some circumstances in which you can't trust them.'

•

Twoday 17 Noctember 2099

Dear Diary,

The Space Hopper has explained the ridiculous geometry of Platterland in terms of an easily comprehensible Flatland analogy. I now see why the angles of triangles here need not add up to 180 degrees, and the right-angled pentagon of what counts for a squarrel in these parts now seems perfectly sensible. I remain vaguely convinced that Platterland's straight lines are really bent, but I now concede that, from a Platterlander's point of view, Flatland's straight lines can be considered bent too. It is dawning on me that what a universe looks like from the inside, and what it looks like from the outside, may be very different.

This, of course, is what the Space Hopper had been telling me for weeks. Stuck-up little twit!

•

'But surely a "straight line" ought to be the shortest path between two points,' said Vikki. 'It is in Flatland.'

'Of course,' said the Space Hopper. 'So?'

'So, I don't see how a *curve* can be the shortest path.'

'Oh dear.' The Space Hopper's customary manic grin turned upside down, from a ∪ to a ∩. 'I thought we'd gone over this already. *They're not curves.* They look like curves to us, because we're on the outside. From inside Platterland, they're *straight.* And – I'm sorry, but they *are* the shortest paths.'

Vikki's head was swimming. 'I hear what you say, but I can't convince my eyes it's true.'

'Let's compare a line that looks straight to *us* with one that looks straight to *them*,' said the Space Hopper. 'What you have to remember is that rulers shrink as they get nearer what we see as the "rim". And our "straight line" lies nearer the rim than theirs. So when you try to measure the two lines, the ruler is shorter along our line than it is along theirs. It's a delicate calculation, but the extra length caused by the shrinkage outweighs the apparent loss of length caused by looking straighter (to us). See it now?'

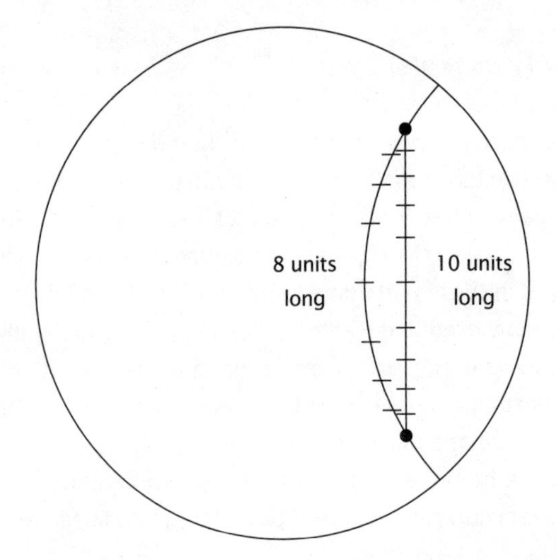

'I suppose so,' said Vikki. 'At any rate, I do see what you're getting at.'

'The relationship between intrinsic Platterland geometry and what we see when we observe their world from orbit is very elegant,' said the Space Hopper. 'We can characterize their straight lines rather neatly. To us, they are arcs of circles that meet the rim at right angles – at both ends. This is called *hyperbolic geometry*.'

'What's it good for?'

'Well, you'll recall me telling you that long ago on Planiturth there was a mathematician called Euclidthegreek. The most influential thing he did was to invent a brilliant way to formalize geometry – in fact, his method can be extended to any part of mathematics. His idea was to start by stating a system of *axioms*: basic assumptions that provide a starting-point for making deductions. You don't have to decide whether axioms are *true*: they determine the "rules of the game". If you dislike the axioms, you just don't play that particular game.

'Anyway, when Euclidthegreek laid down axioms for what in

those days was assumed to be the only possible kind of geometry – in two dimensions, anyway – most of them were pretty simple things: "any two points may be joined by a unique straight line". And "any two lines are either parallel or meet at a unique point" and "all right angles are equal". But one of them was a lot more complicated.'

'What did that one say? Some right angles are more equal than others?'

'No, no. It said, "given any line, and any point not on that line, there is a unique line through that point that is parallel to (meaning 'does not meet') the given line".'

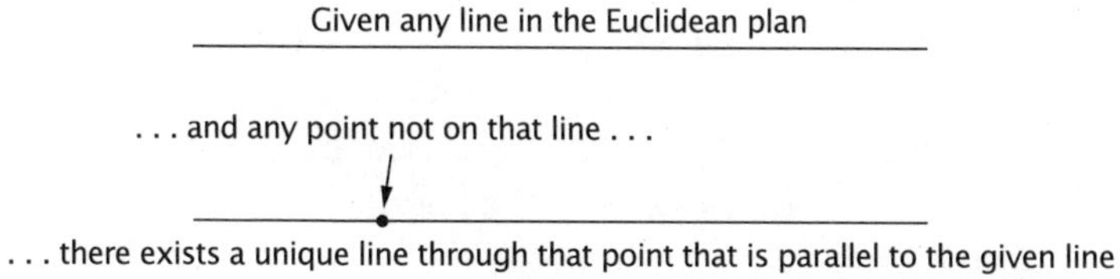

Given any line in the Euclidean plan

. . . and any point not on that line . . .

. . . there exists a unique line through that point that is parallel to the given line

'I see what you mean by "complicated" – but it makes sense.'

'It certainly seems to. And because it was complicated but made sense, people wondered if it could be proved from the other axioms. Because if it could—'

'The complicated axiom would be superfluous,' Vikki finished for him. 'Which would be nice,' she added as an afterthought.

'Exactly. Now, a lot of Planiturthian People tried to do just that. Guess what?'

'No idea.'

'They all failed.'

'OK, so Planiturthians are rotten at geometry.'

'Not at all. Some of them are brilliant. Euclidthegreek was. Mind you, *most* of them are rotten at geometry . . . but that's true about everything. Sports, music, painting . . . But I'm drifting away from the point. Some of the most brilliant Planiturthian mathematicians began to wonder if there was a *reason* why every attempt to prove the parallel axiom failed.'

Vikki thought for a moment. 'But surely there *must* be a way to prove it. I mean . . . why don't you just draw the new line equally far from the original one at every point along its length? That'll do it.'

'Fine. They tried that one. The snag is, how do you know that what you've just described is a straight line?'

'Pardon?'

'How do you know that what you've just descri—'

'I heard you the first time. I just couldn't believe my ends. Of course it is!'

The Space Hopper's uncharacteristic ∩ became even more elongated. 'You think that's a good way to do mathematics? "Of course it is"? What about critical analysis? What about proof? What about logical rigour?'

'Yes, but this is obvious.'

The Space Hopper's ∪ returned, but now it was like the teeth on a shark. 'Really? Well, that's certainly what everyone on Planiturth thought – until they came across hyperbolic geometry. You see, in hyperbolic geometry, all Euclidthegreek's axioms are true—'

'I don't see how they can be, if the lines are curved.'

'I keep telling you, they're *not* curved. But you didn't let me finish. All Euclidthegreek's axioms are true *except* the parallel axiom. And that means you can't prove the parallel axiom from the others.'

'Why not?'

'Because, if you could, the same proof would show that the parallel axiom is valid in hyperbolic geometry. But it's not. Given any hyperbolic-line, and any point not on that hyperbolic-line, there are *infinitely many* hyperbolic-lines through that point that do not meet (which in Euclidean geometry is the same as "are parallel to") the given hyperbolic-line.'

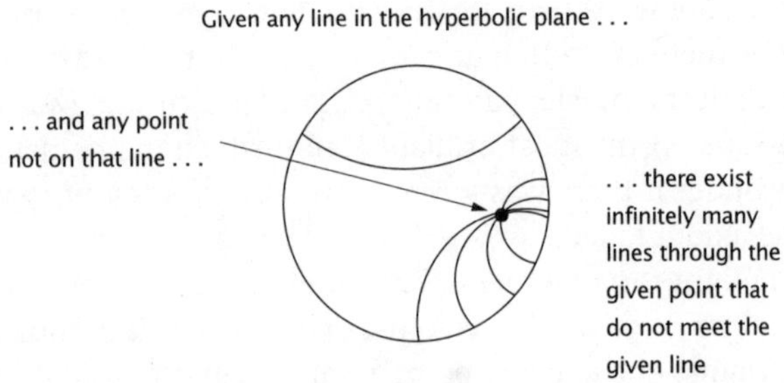

Given any line in the hyperbolic plane . . .

. . . and any point not on that line . . .

. . . there exist infinitely many lines through the given point that do not meet the given line

'It's not about how the real world fits together, is it?' said Vikki. 'It's an abstract statement about the logical consequences of certain assumptions.'

'Exactly. If you spend your entire life in *one* space, you can perhaps be forgiven if you think it's the only space there is, and has the only geometry there is. We Space Hoppers know better: we've hopped between so many different spaces – and now, so do you, even though you're a Flatty with all the usual Flatty prejudices. There are *lots* of spaces and *lots* of geometries, all equally valid in terms of internal logic. And in this particular geometry, hyperbolic geometry, *nearly* all of Euclidthegreek's beloved axioms are valid. Just one isn't: the perplexing parallel axiom. So now you see *why* it's perplexing. *There is no proof.* So there's no point in looking for one.'

'I think I see. It's like trying to prove that the only possible geometry is that of Flatland. But we've seen several different ones already. To us – well, to *me* – the others all seem weird. But if you were used to them, they'd seem entirely natural.'

'Right. You got it.'

Vikki felt pleased. Then a new thought stuck her. 'Space Hopper, what is hyperbolic geometry good for?'

'Good for? What's Flatland good for?'

'It's good for us to live in.'

'OK, so hyperbolic geometry is good for Platterlanders to live in.'

'Yes, I see that. But why should a Flatlander *care*?'

The Space Hopper bounced in excitement. 'Well, one reason – maybe even be the *best* reason – is that Things May Not Be What They Seem.'

'Sorry?'

'You'll see. But not yet. First, you have to understand just *how* different geometries can be. Then, when I've got you softened up to the point where you really *don't* believe your eyes, we can think about the true shape of the Planiturthian universe.' Vikki began to protest, but the Space Hopper was adamant. 'No! Not yet! First, you must learn both patience and imagination.'

Imagination will be easy, she thought. *I've got a good imagination. But patience . . . that's not one of my virtues. In fact, to tell the truth, it's not one of my virtues AT ALL.* But she bit back her protest, since it would only make the Space Hopper's point all the more forcefully.

CAT COUNTRY

Of all the realms of the Mathiverse, there is none stranger than Cat Country. It is (until replaced by something else) the space in which all Planiturthians live – the space *from which they are built*. Yet it is so different from the space that they *perceive* themselves as living in that they find it bizarre in the extreme. Even their scientists find it difficult to understand what goes on in Cat Country. They can *calculate* what's happening, and they can carry out experiments to check their calculations . . . but when it comes to what it all *means*, even the scientists get confused.

You can tell, because if you ask questions about what it all means they stop behaving like scientists and either shout at you or turn into High Priests of some Cosmic Religion. Or both.

Cat Country is ruled by Superpaws, more usually referred to in terms of the owner of its intellectual property rights, an intensely curious Planiturthian named Erwinschrödinger. Not 'curious' in the sense of 'weird': Erwinschrödinger was surprisingly sensible considering what he was interested in. Curious in the sense of 'possessing curiosity'. Now, it is said that 'curiosity killed the cat', but Erwinschrödinger went one better. He killed his cat *and* kept it alive, all at the same time.

Not surprisingly, the poor beast is even more confused than the Planiturthian scientists.

Not surprisingly, the Space Hopper was determined to ensure that Victoria Line became intimately acquainted with Schrödinger's Cat.

•

Suddenly the universe seemed to be *growing*. Not just the space between things, but the things themselves. A piece of gravel that only a moment ago had been lying on the path became a boulder, then a rocky outcrop, then an unorthodoxly shaped mountain . . . Then it became too big to see as a single object at all.

Only the Space Hopper remained his usual size. And Vikki herself, of course.

'Why is everything growing?'

'It's not. We're shrinking.'

The surface of what once had been a pebble kept changing in surprising ways – sometimes rugged, sometimes rippled, sometimes smooth, and sometimes hairy, as if forests were growing on it. Funny clusters of bulbous things appeared, like balloons being blown up, but even they quickly became too big to take in. The pace of shrinkage slowed, then stopped. The surface of whatever it was they were now looking at became sort of lumpy, like dumplings floating in a thin soup. They were small dumplings, and the soup was so thin that it was scarcely distinguishable from empty space. The dumplings formed clusters, like bunches of grapes – but the gaps between the bunches were a lot bigger than you would ever see on a Planiturthian grapevine.

Not far away was a sort-of-beach, itself rather lumpy; and sort-of-waves rolled up towards it, creating iridescent rippled patterns that were hard to make out clearly. Try as she might, Vikki couldn't see where the waves *went*. They just kept coming towards the beach.

There would be time to sort out the waves later. Right now, it was the *lumps* that were bothering her. 'What are those?' Vikki asked. 'Those lumpy things. And why are they so . . . *fuzzy*?'

'Do you mean the round lumps, or the bunches?'

'The round lumps.'

'Those,' said the Space Hopper, 'are atoms. The bunches are molecules.'

Vikki knew about those. 'Atoms are what all matter is made from', she said. 'They're very tiny particles, and they're invisible—' No, *that* couldn't be right. 'Sorry, in*di*visible. And they combine together to form molecules, which is what chemical compounds are made from.' There were atoms and molecules in Flatland, much as in Spaceland, but the differences of dimension made Flatland chemistry a bit simpler than it was in Spaceland. The basic principles, though, were pretty much identical.

'That's what the Planiturthians used to think,' said the Space Hopper. 'Now they know better. Atoms aren't exactly particles, and they're not exactly indivisible. Look, I'll show you.' And he grabbed one of the dumplings and *squeezed*. Out popped three smaller round objects, also fuzzy: two the size of peas, and the other so tiny that only VUE-enhanced senses could see it at all. One pea was green and the other was yellow; the small object was pink. The Space Hopper explained that these were not their true colours – they didn't have true colours. The VUE had just been set to represent them that way.

'Let me just – *bother*!' yelled the Space Hopper. He flapped ineffectually at the yellow pea, sending it skittering through the interatomic soup, in which it momentarily left a faint trail, like ship's wake. They skittered after the errant pea, while the Space Hopper cursed under its breath.

Something small and circular loomed ahead. No, something *large* and circular. Improbably, in this space of fuzzy lumps, it was a perfectly sharply defined object. A bowl. Round the side was written SCHRÖD – no doubt there was more, but it was hidden from view. Inside the bowl was – catfood.

Fuzzy, lumpy, sometimes a bit *wavy* catfood.

The yellow pea plopped into the bowl.

'Good,' said the Space Hopper. 'We'd never have caught it if that hadn't happened.'

'Caught what?'

'The proton.'

'Come again?'

'The yellow pea. It's a subatomic particle called a proton. It's got itself trapped in a potential well. And unless I'm very much mistaken, the electron – the small pink particle – will be sticking to it. They do that, you know. When we first encountered them, those three particles were the parts of a deuterium atom. Now the neutron has been expelled, and the proton and electron have combined to make a hydrogen atom. It's very hard to stop those two joining up if they get the opportunity.'

'What *are* you waffling about?'

'Don't you take that tone with me, young Line. This is very basic stuff – subatomic particles and the Four Forces of—'

'*e*,' a tiny voice interrupted. It seemed to be coming from the pink particle.

'What was that?' said Vikki. 'Did something say "Eee"?'

'No, *e*.'

'Who are you? An electron?'

'Nearly. I am the Charge on the Electron. The fundamental unit of electricity.'

'Oh. I thought the charge for a unit of electricity was about 40p.'

'Not a *financial* charge,' said the Space Hopper. 'An *electric* charge.'

'Well, it would be, for a unit of electricity.'

'I mean, *it's not money*!'

'Oh.'

The Charge on the Electron lost its cool. 'No, not o! *e*! I keep telling you, but you don't listen. I may not be a financial charge, but I do have a value. My value is one ten-quintillionth of a coulomb.'

'That's not very much.'

'No, but there are an awful lot of us, I can tell you.'

•

Threeday 18 Noctember 2099

Diary Dear, you're wondering why the Space Hopper made me spend so much time talking to that funny little Charge creature that lives on an electron.

The answer, in a word, is *particles*.

The answer, in a word, is *waves*.

Aha! I can tell, you're confused! You ask me one question, and I give you two answers. Which contradict each other.

Cat Country is like that.

•

They sat on the seashore of Cat Country, watching the waves surge up the beach – and disappear. Unlike Planiturthian waves, they didn't roll back again.

It was caused, the Space Hopper claimed, by Quantum Tunnelling. Vikki saw no point in asking what that was. She sat, dangling her rear end in the . . . sea . . . and tried to sort out what she hadn't yet understood. 'What *is* an electron, anyway, Hopper?'

'It's a particle.'

'A very, very, very tiny piece of matter?'

'Yes. Smaller than an atom. And it's all in one lump – that's what "particle" means.' The Space Hopper thought about that and corrected himself. 'Used to mean.'

'Isn't *everything* made from particles? What else is there?' asked Vikki.

'Waves. Waves are very strange things. Think of water waves. They seem to be moving along quite fast – and they do roll up beaches, I admit. But if a wave was really a lot of moving water, the land would have flooded long ago. In fact, the water mostly moves up and down – but the hump that People *see* looks as if it's a thing that travels along. However, at each instant, that hump consists of *different* molecules of water. They just trade places. Until they hit the shallows, anyway.'

'But water is just a lot of particles, isn't it?'

'Yes. It's made from hydrogen and oxygen atoms, and those are made from protons, neutrons, and electrons. But water waves are *big*. At very tiny scales, there are waves that *don't* reduce to moving particles.'

'Such as?'

'Light. Light is a wave. Or, at least, that's what Planiturthian scientists had convinced themselves it was. They could even do experiments to prove it. And they thought that electrons were particles – they could do experiments to prove *that*, too.'

'What sort of experiments?'

'I'm sure the Charge on the Electron would be happy to demonstrate a particle experiment', said the Space Hopper.

'Happy? I'd be ecstatic!' said the squeaky voice. 'But I'll need some help. Give me a few minutes to set it up.'

The Charge on the Electron disappeared, and Vikki and the Space Hopper sat beside the quantum seashore watching the waves. The Charge soon returned, accompanied by thousands of identical particles, who were carrying several metal plates and a bottle of oil.

'Idiots! I said *two* plates, not dozens! Oh, never mind, just dump the rest and bring me two. Now, you lot start by lining up the plates, and the rest of you can open up the bottle and create some oil drops. As small as you can manage, please, and be quick about it!'

Soon all was as ordered. 'OK. Now, some of you sit on one plate, and the rest sit on the other one. I'll squirt a drop of oil in between, and you can play roller coaster like you usually do.'

'What on Flat—' Vikki began.

'The other Charges on the Electron are going to take rides on the drops,' the Space Hopper explained. 'That way they transfer electricity from one plate to the other. Now, suppose you could measure the amount of electricity very accurately. What would you expect to find if the electron is a particle?'

'I have no idea.'

'Let me put it this way: each electron carries a charge of one e. So could you detect a transfer of $0.5e$ from one plate to the other? Or $619.74e$?'

'Oh! The amount of electricity transferred always has to be a whole number times e!'

'Spot on, Vikki. And that experiment has been done, and that's the answer you get. If electrons were divisible, you'd get fractions of e, but you don't, so they're not. And that – everyone thought – means that electrons are particles.'

'I'll buy that. What about waves?'

'The important feature of waves is that when they interact, they *add up*. I'll just borrow some of those plates, and you'll see what I mean.' The Space Hopper waded into the sea with a couple of the plates, and set them on end with a small gap between them. Vikki noticed that as the waves encountered the gap, a series of semi-circular ripples formed on the other side.

'That,' said the Space Hopper, rather short of breath from lugging the plates around, 'is a diffraction pattern. It's a sure sign you've got waves.'

'We can *see* the waves,' Vikki pointed out.

'Yes, but *we've* got a VUE. Planiturthian scientists don't have that, but they can detect diffraction patterns even when the waves that make them are invisible. Now, let me add another plate, to make a second gap. To start with, I'll just block off the first gap. What happens?'

'Another series of semicircular ripples, Hopper!'

'As expected. Now I'll unblock the first gap, so that we've got two. *Now* what do you see?'

'Uh – it's rather complicated. A sort of bent chequerboard, that's the only way I can describe it.'

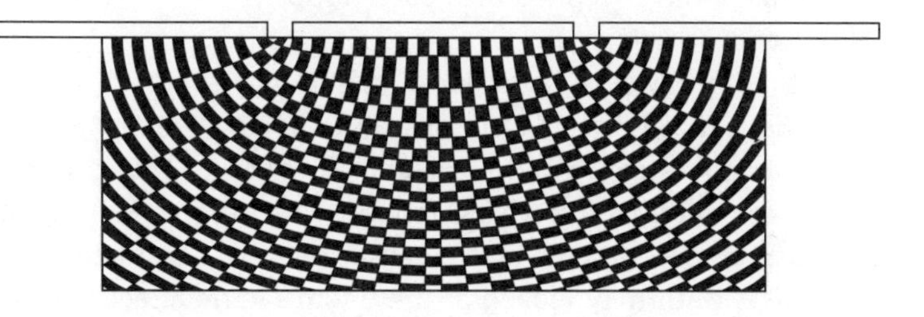

'What's happening is that when a peak of one set of semicircles meets a peak of the other set of semicircles, the two peaks add up to make an even higher peak; and when a trough of one set of semicircles meets a trough of the other set of semicircles, the two troughs add up to make an even *lower* trough. And if a peak meets a trough, they cancel out and the result is in between. That's what I mean by waves adding up. It's called *superposition*.'

'So if we see that sort of pattern—'

'—it's a sure sign of waves. Or, as I said before, so everyone used to think.'

'Used to?'

'On very small scales – down at the sizes we've been shrunk to by the VUE – matter is a bit weird. As Charge on the Electron will now demonstrate.'

'Why *me*? Why not one of the others?'

'You'll do it so *well!*'

'Aw – I *hate* this bit.'

Charge on the Electron stood back, took a run at the middle of the central plate, and hit it with what ought to have been a *splat*. But, Vikki couldn't help noticing, he didn't actually seem to get to the plate – he started to spread out. And then, on the *other* side of the plates—

He formed the bent chequerboard pattern that she had just learned to associate with waves.

'An electron is a wave?'

'Sometimes it feels like it,' said Charge on the Electron in a wavering sort of way. He slowly reassembled into a conventional rounded particle shape.

'It is like a wave,' replied the Space Hopper.

'How can a particle—'

'Behave like a wave? Good question. But it gets worse.'

'I rather expected that,' said Vikki.

'Everyone was convinced that light was a wave. They'd argued about it for hundreds of years, but the diffraction patterns and superposition experiments made it absolutely clear that light really was a wave. Until they looked at the photoelectric effect.' He cast around for inspiration. 'Ah. I knew these plates would come in handy. They're made of a special metal which emits light when . . . Charge on the Electron, will you oblige again, please?'

'Oh, no! I'll be bruises all over . . .' But there was no way out. Again he stood back, took a run at the middle of the plate, and hit it with what this time *was* a pretty convincing SPLAT. He bounced off.

As he did, a thin beam of light shone out of the plate, just for an instant.

'See, Vikki: the metal converts the energy of the impacting electron into light. So what would happen if he hit the plate *harder*?'

'Can't we go for softer?'

'Not yet.'

'You're a pig.'

'We'll get more light?' Vikki hazarded.

'Let's see.'

SPLAT!

'Ow!'

'Yes, you're right. OK, Charge on the Electron – *now* we'll try softer.'

'You rotter!' Splut.

'That's funny,' said Vikki, 'I didn't see any light at all.'

'Me neither,' confirmed the Space Hopper.

'But – we should have seen a fainter beam of light, surely? The lower the energy of the impact, the weaker the light – but it shouldn't just *switch off*.'

'If light is a wave, infinitely divisible, yes. But it did switch off. So what's the explanation?'

'Uh – well, clearly light isn't infinitely divisible.'

'Which means?'

'It's a *particle*?'

'That's what an extraordinarily wise Planiturthian named Alberteinstein was led to conclude. He deduced that light comes in tiny packets with a fixed amount of energy. They're called *photons*. That had already been predicted by a People called Maxplanck. The amount of energy in one packet – one *quantum* – of light is very tiny. It's equal to the frequency multiplied by *Planck's constant*, and that's incredibly small.'

'How small?'

'0.0000000000000000000000000000000006.'

'OK, that's small. Space Hopper?'

'Yes?'

'Is light a wave or a particle?'

'Yes.'

'Both?'

'Sometimes one, sometimes the other. Well – sometimes *behaving* like one, sometimes behaving like the other. Particles and waves are things from classical physics – things we can observe on the human scale. There used to be a thing called the Complementarity Principle that said you could never catch light being both things at once, but that always seemed much too neat and tidy for me. A photon is a quantum object – why should it care which classical object it looks like? And, sure enough, I was right. But on the whole, it tends to resemble one or the other nearly all of the time.'

'And the electron?'

'That's both, too. You can call them wavicles if it makes you feel happier. Though most people now call them quantum wavefunctions.'

'I don't think new names make a great difference. It still seems very weird.'

'That's because we normally live on the classical scale, and we see quantum effects only if we try very hard and use special apparatus. If we'd grown up on the scale of the Planck length, where wavicle physics applies, we'd think that a particle that could *only* be a particle was ridiculously unimaginative.'

'The world is made of wavicles?' Vikki asked, stunned by the newfound vision. *Solid matter wasn't solid.* It was all ripply!

'So the Planiturthians were led to believe. They realized that the wave nature of matter on small scales meant that everything was based on probabilities – on chance. In fact, a quantum wave *is* a wave of probabilities. The places where the wave is biggest – the probability peaks – are the places where the particle is most likely to be observed.'

'Why can't they just observe the entire wave?'

'Because of the Uncertainty Principle, discovered by Wernerheisenberg. If you try to measure the position of a particle, say, then you disturb its quantum wavefunction. In fact, you can't measure a particle's position and velocity at the same instant. Trying to observe one of them changes the other, randomly.'

Weird, though Vikki. Which was a good start. If you don't think quantum theory is weird, then you haven't understood it.

•

Fourday 19 Noctember 2099

The more I look into this particles business, Diary Dear, the more hairy the whole thing gets.

I'll start with *atoms*. When it comes down to it, the Planiturthian physicists have an astonishing habit of choosing really bad names. Like 'Relativity' for a theory where relative motion isn't relative and lightspeed is absolute. But new discoveries have to be given names, and those names have to be based on guesses about what will happen next. If those guesses are wrong, you're stuck with a bad name. I don't *think* that's what happened with Relativity, but it's certainly what happened with 'atom'. Which means *indivisible*, doesn't it, Diary?

Unfortunately, that's about as far from the truth as you can get. There are literally HUNDREDS of *subatomic* particles – *pieces of the indivisible!!!* Proton, neutron, electron, photon . . . but those are just the best known. There are muons and tauons and neutrinos and pions and all sorts of others . . .

The list of fundamental particles is HUGE. It's hard to see how they can be fundamental when there are so many of them. But the irony of it is – so the Space Hopper says – that the best way to reduce the number of fundamental particles is to introduce a lot of extra ones.

This sounds mad, but it works because the old ones can be built up from the new ones. So when it comes to 'fundamental building blocks' you can keep to a fairly short, and very structured, list. Sort of sub-subatomic, then, if you catch my drift, Diary.

The most important of these sub-subatomic particles are known as QUARKS. (The name rhymes with 'corks', not 'marks', OK?) Protons and neutrons, for instance, are made by sticking quarks together. The proton is two *up* quarks plus one *down* quark; the neutron is two *down* quarks plus one *up* quark. The omega-minus particle – no, I dunno either – is three *strange* quarks.

It's like this: there are six 'flavours' of quark: *up*, *down*, *top*, *bottom*, *strange*, and *charmed*.

Quarks always come either in threes, or as a quark–antiquark pair. Mesons are like that, for instance. You never observe a solitary quark. How do they know that? Because quarks have charges that are ⅓ or ⅔ the charge on the electron. So if they occurred on their own, they'd stick out like a sore vertex.

•

Where there is catfood, there will be a cat. (Often the same one.)

Vikki looked anxiously around, but the animal was nowhere to be seen. She walked round the bowl of catfood. On the other side, it read 'INGER'.

'Why does the bowl say INGERSCHRÖD?'

'No, no, Vikki: that reads SCHRÖDINGER.'

'Manufacturer?'

'No.'

'Catfood advert?'

'No.'

'Logo?'

'No.'

'The cat's called Schrödinger?'

'No.'

'Oh.'

'Maybe the cat dings schrös.'

'Dings?'

'Rings like a bell.'

'Schrös?'

'Rather like shrews, I think.'

'Oh.' The conversation lapsed into an awkward silence. Finally, Vikki said, 'Space Hopper, you made that up! You have no idea what the name means, and there's no such thing as a schrö!'

'Don't complain to me,' said the Space Hopper. 'Actually, I do know where the name comes from, but here's the cat himself. You can ask *him*.'

'Where? I don't see him.'

'He's starting to materialize. Look over there – you can see the glum.'

'The *glum*? What's a glum?'

'The opposite of a grin. This is a very sad cat.'

A ∩ shape had formed in mid-air, near the bowl. It slowly solidified into a doleful and rather fat grey-and-white striped cat, which sat down by the bowl and began to eat from it.

'Excuse me—' Vikki began.

'Don't mention it,' said the cat in a mournful voice, and carried on eating.

'No, I mean I'm sorry to interrupt you—'

'Then don't!' snapped the cat.

'I *did* say "excuse me",' Vikki pointed out.

'And I told you not to mention it. Which you've just done. Again. Don't you ever *listen*?'

'You're a very strange cat,' said Vikki.

The cat stopped eating and stared at her. 'Look: you're an *observer*, right?'

'Sorry, what do you—'

'You're *observing* me, yes?'

'Well, if you put it that way. I suppose I am.'

'Then I can't be a strange cat,' said the cat. 'I'm only a strange cat when you *don't* observe me. When you're not looking, I'm totally weird. But you'll never *observe* me being weird. To you, I'll always look normal.'

Vikki was prepared to dispute that – she found the creature both fascinating and irritating. But politeness was a virtue to be cultivated. 'Are you hungry?' she asked.

'That depends,' said the cat.

'On what?'

'On whether you're observing whether I'm hungry.'

'I'm observing you *eating*', said Vikki.

'That's different,' said the cat. 'Inferences don't count. As long as you're not actually *observing* my state of hunger, then I could be hungry, and I could be bloated with food and desperately stuffing the last morsel down my throat. Or I could be half of each at the same time.'

Aarrrgggh! Why was everyone in the Mathiverse like this? 'Space Hopper, I need your help. This cat isn't making any sense.'

The Space Hopper wandered over.

'That's because you haven't heard my tale,' the cat said.

'Don't be silly, you *see* a tail, you don't hear it. And I can see your tail perfectly well.'

'I'm a bit sensitive about it,' said the cat.

'Don't worry, Mr Cat – I'm not going to tread on it.'

'My name,' said the cat, 'is Superpaws.'

'That's a funny name.'

'What's *your* name?'

'Vikki.'

'And you say *mine* is funny. Ha! Be warned, my tale is a sad one.'

'No it's not! It's stripy and twirly and really quite beautiful!'

The Space Hopper took Vikki to one side and gently corrected her misapprehension. Then they sat down while the cat related its story. (*And if you'd* called *it a story in the first place, Mr Superpaws, none of this embarrassment would have happened!* But she was too polite to say so.)

'For many years,' the cat began, 'my life was uneventful. Until one fateful day I saw a job advertised in an experimental laboratory. I should have realized. "Cat" and "experimental laboratory" – it's a disturbing juxtaposition. But being a bit short of readies, you understand, I took the plunge and sent in my application. To my delight – well, it was delight to begin with – I secured the position.'

'Which was?' Vikki enquired.

'I had to sit in a box. Now, sitting in a box is something cats are pretty good at. "Easy-peasy", I thought to myself. What they didn't tell me was that I'd be shut up in the box with some kind of gadgetry stuff, and I couldn't get out again until some geezer opened the box. All this was before the Animal Rights Movement, you see. Nowadays it would never be allowed, but back then, nobody cared.

'So, I'm sitting in this box, and I can't get out until this geezer lets me out, and I look around to see what's there – being a cat I can see in the dark, and anyway, there was this lump of stuff that was *glowing*, so I thought, "That's nice, they've left me a nightlight." And there was some kind of glass tube, with nothing in it. Well, it *looked* like nothing – I only found out the real story later. The glass was a bit fragile-looking, and then I noticed this dirty great hammer on a hinge, held up by a length of very thin thread. So I thought, "Someone ought to get rid of that hammer before it falls down and smashes the glass." There was a pair of scissors, opened round the thread, and there were lots of coils and springs and gear-wheels and things running from the handles of the scissors to some incomprehensible gizmo that was pointing at the glowing stuff.

'Since it made no sense at all, I went to sleep, like any normal cat. Next thing I know, the lid comes off, and someone's poking their nose inside. So I lie low, hoping to get a bit more shuteye.

'"Oh, look, it's dead!" says a voice. 'Course, I knew I wasn't – we cats can tell that kind of thing – but I wasn't sayin' nuffin'. "The detector must have registered the decay of a radioactive atom and released the poison gas," says another.

'*Poison gas!* Well, I was ready to leap out of that box like a shot, I can tell you. But then the first geezer says something that pins my ears back. "So his wavefunction collapsed, then?" Look, matey, I'm *proud* of my wavefunction, and I'd never let it collapse even if you paid me huge bundles of dosh to neglect it. I wash my wavefunction every month, whether it needs it or not. Anyway, "Yes," says the second geezer. "He must have been in a superposition of states –

part alive, part dead. Then we opened the box, and observed him – and collapsed his wavefunction to 'dead'."

'Well, that did it. I opened one eye, then the other, and glared at them. And do you know what the first geezer said?'

'No.'

'"Perhaps we should have used stronger poison." Well, *really!*' The cat stopped for a moment to lick its paw and wipe it over one ear. 'Later, I figured it out,' he continued. 'I was a quantum cat, see? Subject to the Principle of Superpawsition. Hence the name, of course. Like a wave, I could interfere with myself, if I wanted to – though I hasten to add that I'm not that way inclined *at all.*

'When the box was shut, the two geezers outside couldn't observe what was happening, so they had to assume that the radioactive atom was in a quantum superposition of the states "decayed" and "not decayed". Which, bearing in mind the detector, hammer, glass vial, and poison gas, indicated to them that *I* was in a quantum superposition of the states "dead" and "not dead". When they opened the box and *observed* my state, however, their actions immediately collapsed my wavefunction. Luckily for me, it plumped for "not dead", and I was able to hoof it before they shut the lid and tried again. But they thought I *was* dead. Too busy observing *me* to notice that the hammer hadn't moved.'

'That's awful,' said Vikki.

'The worst is yet to come,' said the cat. 'It so happened that just as I escaped from the box, their attention was distracted by discovering that the poison vial was still intact, and they didn't actually *observe* me escaping. So all they could assume was that I was in some quantum superposition of the states "escaped" and "not escaped". And I've been wandering through Cat Country ever since, wondering exactly when my wavefunction will collapse – and whether, when it does, I'll find I've been in the box all along, when I *thought* I'd eluded their Evil Clutches.'

•

'Let's see if I've got this straight,' said the Space Hopper. 'For the purposes of physics, you are not so much a cat as an extraordinarily large system of fundamental particles, all interacting in some incredibly complicated way. Since each individual particle is a quantum wave, it has a quantum wavefunction. Therefore *you* have a quantum wavefunction—'

'It's more that I *am* a quantum wavefunction,' Superpaws interrupted.

'—which is therefore subject to the Principle of Superpawsition . . . I mean, Superposition. Your quantum wavefunction, when unobserved, can be – indeed, most likely is – in a superposition of states.'

'That's the gist of it,' said the cat.

'When you are sealed up in your impermeable box, then it is impossible for anyone to observe you, so—'

'Wait a minute,' said Vikki. 'Why can't Superpaws observe *himself*? Isn't he an observer . . . inside the box?'

'I wondered about that myself,' said the cat, 'but what they all say is that because I can't communicate my observation to anyone outside the box, it doesn't count.'

'That seems very peculiar to me—' Vikki began, but the Space Hopper was anxious to continue with his summary.

'—since it's impossible for anyone *else* to observe you while you're inside the box, your wavefunction can remain in a superposition of states.'

'Yes.'

'The relevant states being "alive" and "dead".'

'Yes.'

'Why?' asked Vikki. 'Why couldn't Superpaws's state be a superposition of "hairy" and "hungry"?'

'Dunno,' said the cat. 'To be honest, that's my most probable state – what physicists call the "ground state".'

'Or "covered in purple spots" and "descending from a tall building by parachute", for that matter,' said Vikki, warming to the task.

'Not *quite* as likely,' said the cat. The Space Hopper felt a kind of *itch* at the back of his mind, a vague memory that wouldn't pop into conscious view.

'No, but in the quantum world, anything that's remotely *possible* has to be taken into account, yes?' Vikki persisted.

'Yes,' said the cat, 'that's what worries me.'

'Now, let me ask you about observations,' said Vikki. 'Just what constitutes an observation in quantum theory?'

'It's any interaction between the system being observed and its environment that generates a well-defined *number*,' said the Space Hopper. What *was* that elusive thought?

'So if I count Superpaws's tail and get "one", that's an observation?'

'Yes. If you count Superpaws's tail and get *five*, that's an observation, too. But not a very good one.'

'And what happens when someone makes an observation?'

'The wavefunction collapses to some pure state.'

'What's a pure state?'

'One that can't be described as a superposition of other states. Which, by some mathematical trickery, is the same as one that leads to a numerical answer. So pure states are the *only* ones that can be observed.'

Vikki thought about that. 'And "dead" is a pure state?'

'Yes, because you can observe it.'

'I'm not sure that isn't circular logic,' objected Vikki. 'What number do you get for a dead cat?'

'Zero. Breaths per year.'

'Hmm. I'm not convinced. Is "dead" really a quantum state?'

'*Everything* is really a quantum state.'

'No, you didn't hear the emphasis. Is it really *one* quantum state?'

'Sorry?'

'Couldn't a cat be dead in more than one way? I mean – run over by a bus, eaten by an alligator, squashed by a rampaging hippopotamus—'

'Do you *mind?*' protested the cat. 'I think you *enjoyed* that!'

'Sorry, just listing hypotheticals. How can you tell from the quantum state of a cat that it's dead?'

'I don't know.'

'Alive?'

'Same problem. I'm not sure you can even tell it's a cat.'

'So if "dead" is really lots of possible states, and "alive" is really lots of possible states, what happens when you superpose half of each? You get even more possibilities. For all we know, "covered in purple spots and descending from a tall building by parachute" is 13 per cent dead plus 87 per cent alive if you choose the right "dead" and "alive" states. I mean, suppose it's "only just dead and covered in *lots* of purple spots" and "extremely alive and vibrant and descending from a very tall building by an oversized parachute"? Wouldn't 13 per cent dead plus 87 per cent alive work out as "covered in purple spots and descending from a tall building by parachute"?'

'I have no idea. But your general point is a good one.'

'Superpaws, it's all nonsense, you know. You're a *cat*, not a wavefunction. "Cat" is a classical concept, not a quantum one. I suppose that at any given moment you've *got* a quantum wavefunction . . . no, even that's not really true.'

'Why not?' asked the Space Hopper.

'Because he's interacting with his environment. So you can't split off the part of the quantum wavefunction that's *him*, and the part of the quantum wavefunction that's the environment. For instance, are the molecules in the food that have gone into his stomach part of *him*, or part of his environment?'

'I never thought about that,' said the cat. 'At first they're not part of me, and later they are. The main thing is to get as many of them as I can. But this business about the collapse of the wavefunction: it definitely works, you know, they've done experiments!'

'With a cat?'

'Well, like I told you, they were working on it when I . . . No, with electrons.'

'An electron is a very simple quantum system, Superpaws. A cat isn't. It's *immensely* complicated. So terms like "pure state" and "superposition" aren't terribly clear-cut, are they?'

The Space Hopper finally remembered what had been bugging him. 'Vikki, you're right. Superpaws has a very short decoherence time.'

'Blimey,' said the cat, 'now I'm feeling *really* safe.'

'Superpositions of pure states kind of fuzz out in systems composed of large numbers of particles. Their phases decohere.'

'Deco here, deco there – whatever,' said the cat. 'I can't abide jargon.'

'OK. Er . . . quantum states are usually interpreted as probabilities. If you observe a mixture of pure states, say 13 per cent dead plus 87 per cent alive, that is interpreted as a probability of 13 per cent that you observe a dead state, and 87 per cent that you observe an alive state.'

'Well, at least the odds are on my side,' remarked the cat. 'I'd prefer better ones, though. Those are about the same as Russian roulette.'

'However, the probability is just the *amplitude* of the state – the *size* of the wave that corresponds to it, so to speak. As well as an amplitude, every quantum wavefunction also has a phase.'

'I've been going through a phase,' said the cat. 'Didn't know I *had* one, though.'

'The phase is how far through the cycle of wave-motion the state has got at a given moment,' said the Space Hopper. 'Relative to some specific choice.' He was struggling. 'I can show you the formula if you—'

'You mean, like – the wave goes up and down and up and down, over and over again, and the phase is whether it's up, down, or somewhere in between?'

'Thank you, Vikki, that's exactly what I mean. Though "up" and "down" must be taken metaphorically. Now, the number you get when you observe a state *doesn't depend on the phase*. But the wavefunction you get by superposing states *does depend on the phase*. People used to think you couldn't observe the phase, but now they just think it's extraordinarily difficult.

'Anyway, every quantum system undergoes complicated changes in phases as it interacts with its environment. That's "decoherence". If it's a simple system like an electron, it can maintain its phases in much the same relationship for quite a while, so "superposition" has a well-defined meaning, and it can behave like a quantum particle. But the decoherence time – how long it takes for the phases to get jumbled up – increases very rapidly as soon as the number of particles increases. Superpaws's wavefunction would decohere in a time so short that it doesn't even make sense to speak about it in quantum theory.'

'Wonderful!' said Vikki.

'What good is all that to me?' asked the cat.

'It means, my dear Superpaws, that if you were ever in a state that was a superposition of "dead" and "alive", you would have stayed in that state for such an incredibly tiny fraction of a second that no one could ever have caught you in it. Because you are a large quantum system, interacting with your environment, you don't *behave* like a quantum system at all. You behave like a classical one.'

The cat digested this information. 'But the box stopped me from interacting with my environment.'

'No, there was more than enough environment in the box. The interaction of the radioactive atom with the *detector* made the wavefunction decohere – you weren't even involved at that point.'

'I would have been,' said the cat glumly, 'if the detector had registered the decay of a particle. Still, I see what you're telling me. So I was never in a superpawsition of "dead' and "alive"?'

'Not one that lasted long enough to count.'

'What state *was* I in, then?'

'You were *either* dead, *or* alive,' said the Space Hopper. '*You* would know which – at least if it was "alive". An external observer *wouldn't* know, though – not until they opened the box. But there's nothing mysterious about that! Nearly everything that happens to us is like that. We can't predict *what* it will be, we just have to wait to find out. That's not quantum. It's just ignorance.'

There was a long pause. Then, 'I can live with that,' said Superpaws. 'It's my stock in trade.' And his face lit up with a ∪, almost as broad as the Space Hopper's.

•

Fourday 19 Noctember 2099 [continued]

Something's been bothering me for a while, Diary my Friend, and I finally managed to put my endpoint on it.

You know that the Mathiverse and the Planiturthian universe are different, yet strangely intertwined. The Mathiverse is a collective mental creation of Planiturthian intelligences – yet the structure and behaviour of the Planiturthian universe is somehow governed by the Mathiverse. So have these bizarre creatures *invented their own universe* out of pure mentality?

I asked the Space Hopper. And he laughed.

After a while, he apologized and said it was a really good question, but he found it amusing because some Planiturthian scientists and philosophers believe just that. They think that unless at least one Planiturthian Mind is observing the universe – to 'collapse its wavefunction' – then it doesn't exist.

Most, though, think this is rubbish. ('Fortunately', he added as an afterthought.) A universe is even more complex than a cat, so its decoherence time must be even shorter. And anyway, how could the brain to house a Planiturthian mind ever have evolved in a nonexistent universe?

Good point, Diary.

What WAS going on, then? I asked him.

He pointed out that Planiturthians don't possess a VUE. They can't *experience* the Mathiverse directly, like I'm doing. But that, he said, is because *they* are real whereas *I* am a Virtual Unreality Construct. I got annoyed, but he says he's one too.

So how *do* Planiturthians experience the Mathiverse, then? Instead of a VUE, they use an IMAGER. Imagination, Mathematics, Analogy, Generalization, Extrapolation, and Recursion.

Take, for instance, their conception of space. They look around and imagine themselves to be on a flat plane with an extra up/down direction. (Planiturth is round, but no matter: it's also BIG.) They turn this into the mathematics of 3D Euclidean space. They use analogies with 2D space, where they can draw pictures, to understand 3D. But – and this is crucial to understanding how their minds work – *they don't stop there.* They generalize to 4D, 5D, *n*D, ∞D. And they extrapolate concepts from their 3D experiences, so planes extrapolate to hyperplanes, spheres to *n*-spheres, and so on. Finally (only you'll see it's *not* final at all) they recursively go back to the imagination stage and start all over again. Which is what's led them from a simple 3D Euclidean spatial model to the far more sophisticated and intricate geometry of the quantum universe.

'So it's all in their minds?' I asked the Space Hopper.

'No,' he said. 'Their minds are all in It. Their brains are built from Planiturthian-universe matter, obeying Planiturthian-universe rules. Their minds are processes that go on inside their brains – and many of those processes are internal representations of that external universe. So, not surprisingly, Planiturthian minds construct – by the collective use of IMAGER – a Mathiverse that mimics the effects of those external rules pretty well. They don't always get the rules *right* – for all they know there may not be any ultimate rules at all – but they keep tinkering, and slowly the correspondence between Mathiversian rules and observed reality becomes extraordinarily accurate.'

'So that's all there is to it, then?' I asked him.

'Not quite,' he said. 'There's something very mysterious going on, too. Why is it that Planiturthians live in a universe where IMAGER works, anyway?'

'Why?' I asked him, after a long silence.

'Beats me,' he said.

12

THE PARADOX TWINS

'Grosvenor?'

'Yes, Lee?'

'You really miss her, don't you?'

Grosvenor Square dropped his newsscroll and glared at his wife. 'Who?'

'Vikki.'

Grosvenor went dark grey with repressed anger. 'I'm surprised you can bring yourself to utter that name in this house after what she's done to us!' Jubilee said nothing, but she was clearly on the verge of tears. 'Not even a postcard or a phone call . . . that's what I find really annoying, Lee. I might forgive her running away – she's an adult, when it comes down to it. But not like this.' He shook his vertices in disbelief. 'This just leaves us stuck in emotional limbo. It's so *unfair*!'

'It's not like our Vikki, is it?'

'Not like her at all.' He took a deep breath. 'I'm very worried, Lee. Maybe . . . maybe somebody's stopping her from getting in touch. Maybe she's in some kind of trouble. I – I know I don't talk about it much, dear, but it would just upset everyone. So I keep my feelings to myself.'

'And that's what upsets everyone, Grosvenor. Don't you understand that? You don't need to hide your feelings from *me*. How do you imagine *I* feel?'

'I – I know . . . but even so, I think it's sometimes more upsetting talking about her, when we can't *do* anything about it. I just . . . wish . . . she'd get in touch. It would set my mind at rest to know that somewhere on Flatland she's safe and sound, and that she's put all this three dimensions nonsense behind her.'

•

Fiveday 20 Noctember 2099

OK, Diary: now get THIS!

You'll remember I told you that the Mathiverse is both a Planiturthian construct and the driving force of the Planiturthian Universe. Yes, I *know* it's mad, but that's Planiturthians for you . . .

Anyway, a consequence of this loopy self-referential nature of Planiturthian civilization is that, uniquely among the inhabitants of the Mathiverse, they keep changing their minds about which Space they are actually *in*.

In his *Romance of Many Dimensions*, my great-great-granddad Albert uses the term 'Spaceland' for the geometry of Planiturth. I mean, here's how his memoirs START:

> *I call our world Flatland, not because we call it so, but to make its nature clearer to you, my happy readers, who are privileged to live in Space.*

And he talks about Spaceland sailors, and Spaceland children, and a number of other things which – thanks to some useful guidance from the Space Hopper – make it clear to me that old Albert managed to confuse Planiturth with a Mathiversian Construct. No surprise, then, to find that Spaceland was the Mathiversian Construct that was believed to be a valid geometry for REAL Planiturthian space in the days when Albert visited what he was TOLD was Spaceland. (Remember, Diary Darling, that for profound reasons of Cosmic Synchronicity, deriving from Narrative Imperative, the enumeration of years in the Planiturthian Universe proceeds in step with that in Flatland, except for a numerical difference of 100, so that our 1999 is their 1899, and our 2100 is their 2000.)

Well, as you have no doubt anticipated, by the time Yours Truly got to visit Planiturth the fickle creatures had changed their minds. No longer did they imagine their world to be an ideal 3D

Euclidean Space. Not a bit of it. It had become a 4D Spacetime. With various bells and whistles which I shall reveal at a later date!! They had come to this momentous conclusion, apparently, because of an Albert of their own . . . But I am getting ahead of my story, as usual. I am *so* impatient, Diary Dear, to tell you the astonishing things I am discovering.

•

Vikki and the Space Hopper were sitting in a bus.

The Space Hopper had bought them two tickets for a kind of historical package tour of Planiturthian Physics. Actually, it was an omnibus, meaning that it could go anywhere. Not any-where on Planiturth, but *anywhere*. It was a Universal Touring Machine. And in the wild expanses of the Mathiverse there's an awful lot of anywhere to go, but most of it is so bizarre that it's probably best not to. The trouble with package tours, though, is that once you've paid your deposit you don't have much of a say in where you go.

The bus was chugging its laborious way along Continuum Carriageway, a metaspatial bypass that avoided the need to drive through Topologica, which was always a nightmare because nothing ever stayed the same shape. The driver was desperately looking for the exit to Pyramid Park and hoping he hadn't passed it by mistake, when Vikki and the Space Hopper heard the sirens screeching in the distance, far behind them. The driver must have heard them too, because he pulled over to the hard shoulder and killed the bus's motor.

WHEEEEEEEEEEEEEEEEEEEEOOOOOoooooooooooooooooooooo

Whatever it was shot past so quickly that they never caught a glimpse of it.

'What was *that*?' screeched Vikki.

'Dunno,' said the Space Hopper. 'I'll have a word with the driver.' He hopped along to the front of the bus and engaged in animated conversation with the person behind the wheel. After a few minutes, he hopped back again. 'Um. Let me see if I can summarize . . . you've heard of the *Fire* Brigade?'

'Of course,' said Vikki. Every city in Flatland had its own Fire Brigade. 'They ride round in fire engines and put out fires. Try to, anyway.'

'That was the Light Brigade.'

'They . . . ride round in light engines and . . . put out lights?'

'Mmm . . . sort of,' said the Space Hopper. 'They'd put out the light if they could catch it, definitely – but they've never succeeded. They *chase* light. But it's a wild-goose chase, because in this Spacetime they have no chance of succeeding. Of course that doesn't stop them – it's their *job*, you see. And they're very conscientious, the Light Brigade.'

'What a shame. Why can't they catch the light?'

'Because it's faster than them.'

'I think,' said Vikki, concentrating hard, 'that's just another way of saying they can't catch it.'

'I guess. The real point is, it's faster than *anything* – except itself. It's the fastest thing there is. Didn't you notice all those round signs along the roadside that say ©?'

'I thought they were copyright notices.'

'No, *c* is the speed of light, 300,000 kilometres per second. Those © signs tell us the universal speed limit in this part of the Mathiverse – the Relativistic Spacetime Continuum.'

Vikki thought about this. 'I didn't realize light had a speed. When you switch on a lamp, you don't see the light seeping across the room towards you.'

'Vikki, even if it was doing that – *what would you see it with*?'

'Light – oh. You only see it when the light arrives, of course. Sorry.'

'There's more to it than that,' said the Space Hopper. 'You don't *hear* the words other people are speaking making their way across the room towards your ears, either. For the same reason. But sound takes a perceptible time to travel from one place to another. That's what makes echoes work – you hear your own words bouncing back to you *with a time delay*. Caused, of course, by the time it takes the sound to get to whatever it bounces off and come back to your ear. With me so far?'

'Yes. But why aren't we aware of echoes in a room?'

'In a *big* room we do. Even in a small room, echoes give the sound its character. But mostly we don't notice because the time it takes sound to cross a room is a tiny fraction of a second.

'Light is similar – but it's a *lot* faster. You see a flash of lightning almost the instant it happens, but the thunder takes time to reach you, and you're aware of a delay. About three seconds per kilometre.'

'I always wondered why that happened.'

'As far as lightning goes, light *could* be instantaneous. Infinitely fast. But it's not. Planiturthians found ways to measure it – astronomical ones, to begin with – and they found that it travels at the aforementioned 300,000 kilometres per second. Our old friend Alberteinstein – yes, the People who realized that light can behave like a particle – used that one simple fact as the basis of an entire theory of Spacetime. He called his theory "Relativity".'

'Because it said that everything was relative?'

'Pretty much the exact opposite! The main thing to remember about Relativity,' said the Space Hopper, 'is that it's an extraordinarily silly name.'

'Then why use it?'

'Historical accident,' said the Space Hopper. 'The Planiturthians are stuck with it. The whole point of Relativity is *not* that "everything is relative" but that one particular thing – the speed of light – is unexpectedly *absolute*. Here, take this gun.' A small pistol had materialized beside them.

'*Gun?*' Vikki squealed. 'What would I do with a gun?'

'It's all right, it's a potato gun. Point it forwards and fire it.'

'But I might hit the driver—'

'No, aim at that big sign beside the road, the one that says TIREDNESS CAN KILL. Oh, and you'd better load it first.' The Space Hopper passed her a potato. She looked baffled. 'Just put it in the chamber there. Like I said, it's a potato gun. It fires potatoes. Whole ones.'

Vikki had got used to strange requests from the Space Hopper, so she loaded the gun with a medium-sized King Edward and took steady aim, just past the driver's left ear. She began to *squeeze* the trigger . . .

'Wait! Hold it! First, a question for you. You're travelling in this bus at 50 kph. A stationary gun fires a potato at 500 kph. How fast will this potato travel?'

'Um . . . Well, it gets a 50 kph boost from the bus, so I guess you add the speeds up, like Running Turtle would have done. That makes 550 kph.'

The Space Hopper pulled a remote velocimeter from the voluminous folds of metaspace. 'We can measure the speed with this. Right, fire away, and we'll see.'

The gun went off with a bang. The driver ducked as a potato-shaped hole appeared in the windscreen next to his head, and the bus wobbled all over the metaspatial road. The driver turned and glared at them, then turned back again quickly before he hit a seven-dimensional truck.

'Sorry,' said the Space Hopper. 'Physics experiment.' The driver muttered something unintelligible and probably obscene. The Space Hopper consulted the remote velocimeter. 'Well, it's 549.999999999 something. Good enough for government work.' He handed Vikki a torch. 'Now, let's try it again with light.'

'Light?'

'Light travels at approximately 1,080,000,000 kph (300,000 kps). But the bus is moving at 50 kph. So at what speed does the light hit the sign? Not *that* sign, we've passed it. The one coming up next.'

Vikki thought about this. 'I guess the speeds add up, like they did for the potato. So that ought to give a speed of 1,080,000,050 kph.'

'OK, let's try it. Ready . . . aim . . . *switch*!'

A pool of light illuminated the sign, before the bus chugged past.

'Too quick for me to tell,' said Vikki.

'Too quick for anyone. But the remote velocimeter knows – it's extremely accurate. And it says the speed was 1,080,000,000 kph.'

'What happened to the extra 50?'

'Not there. Extraordinary, isn't it? If you fire light from a moving vehicle, the light travels at exactly the same speed as if the vehicle wasn't moving at all.'

'That's mad. Surely relative velocities don't—'

'In Spaceland geometry, no. In Relativistic geometry, yes. And to make it worse, even though our bus is moving, the speed we measure for light passing along the bus is exactly the same as the speed that would be measured by a stationary observer.'

'Are you sure?'

'Let's ask one.' The Space Hopper leaned out of the window and shouted to a motorist trying to change a flat torus. 'Oi, mate! What speed do you measure for light?'

The motorist made a rude gesture.

'Picked the wrong observer, I think,' said the Space Hopper, unabashed. 'Ah, here's a policeman. I'll ask him. Driver – slow down! Stop for a moment, there's a good chap! Pardon me, Officer, but what speed do you measure for light?'

'1,080,000,000 kph,' said the policeman. 'It's the lawful speed limit, sir – you must have seen the signs. *Nothing* goes faster than 1,080,000,000 kph. At least, if it does, it'll have me to reckon with. *And* a speeding ticket.'

Vikki still felt that there was a difficulty. 'Look, Hopper, I know I'm probably being unimaginative, but you can't limit the speed of light just by passing a speeding law. The only law that can limit the speed of light is a law of physics. And physics says that relative speeds add up.'

The Space Hopper shook his horns. 'Ah. Physics prior to Alberteinstein said that, certainly. But not after. You can do very sophisticated experiments, with highly accurate equipment, and *you get the same answer* – as a People called Michelsonandmorley discovered between 1881 and 1894, Planiturthian time. He was trying to detect the motion of Planiturth relative to the "aether", an all-pervading fluid that was thought to transmit all electromagnetic radiation, light included. If Spaceland physics were correct, that motion should show up as a difference in the apparent speed of light when Planiturth was at opposite points of its orbit, moving in opposite directions. But Michelsonandmorley couldn't find any difference in the speed at all, even with very sensitive equipment.'

'All that proves is that Planiturth must carry the aether along with it when it moves in its orbit,' Vikki objected.

That had never occurred to the Space Hopper, and for a moment it threw him. 'That's quite a cute theory, Vikki. But it's wrong, because . . .' he searched desperately for a counter-argument '. . . you'd expect to see funny effects in the light from distant stars if the aether was swirling around like that. Michelsonandmorley concluded that either there wasn't an aether at all, or Planiturth *wasn't* moving relative to it – which he didn't think was credible – or that there was something pretty weird about light.'

•

The bus resumed its journey, and the Space Hopper continued with his explanation. 'In 1905 Alberteinstein turned Michelsonand-morley's observation into a theory – Special Relativity, it was called – to the effect that there *is* something pretty weird about light. And he pointed out that this meant there had to be something pretty weird about space, too. Well actually he wasn't the first to do that, but he was the first to understand the Big Picture behind it all.'

'How can space be weird?'

'Let me adjust your VUE to exaggerate the effect I'm talking about. It will slow light down to a modest running pace. Now, you sit on that seat, and watch me hopping past you. See if you notice anything different about me.' The Space Hopper bounced past, very slowly.

'Not a thing. You look perfectly normal to me.' Vikki coughed and corrected herself. 'As normal as you ever do, that is.'

'Good. That's what I *should* look like. But now I'll speed up.'

Again the Space Hopper bounced past her, getting some aggrieved looks from the other passengers.

'You look – Space Hopper, you look *thinner*.'

'That's right. And so do you. Not *thinner*, though: you're a line and your thickness is zero anyway. Shorter. But when you say I look "thinner", what you really mean is that my width looks shorter, yes?'

'Yes.' Vikki curled her tip round to look along her length. 'But I'm the same length as usual.'

'To *you*, yes. To me, too – while I'm sitting here and not moving. But if I move fast enough, you look shrunken to me.'

'But—'

'I know, you don't look shrunken to you. That's because you're not moving relative to yourself. But you *are* moving relative to me, so to me you look shorter.'

'But . . . wait. If I'm moving relative to you, then you're moving relative to me. So to me, *you* look shorter.'

'Correct! It's perfectly symmetric! When we are in relative motion, I look shorter to you, and you look shorter to me. Whereas we each look our normal length to ourselves. Isn't that amazing?'

'Astonishing. I don't believe it.'

'It's a necessary consequence of light having an absolute speed.

Light from different parts of a moving object reach you at different times, having travelled different distances. Combined with an absolute speed for light, that makes the moving object seem to *contract* in the direction of its motion. Alberteinstein realized that this contraction was a *real* physical effect. A lot of other Planit-urthians, like Hendriklorentz and Henripoincaré, were working on the same idea, but they kind of saw it as a mathematical fiction. Not so old Alberteinstein. He was made of sterner stuff!

'And it doesn't end there,' continued the Space Hopper. 'Time has to slow down when you move quickly, too.'

'Time usually seems to speed up when you move quickly, in my experience.'

'Ah, that's psychological time. I mean time as measured by a clock.'

'Oh. But why does time—'

'I'll bounce past you again, but this time I'll shine a torch. Still with artificially slow light, OK? I'll give you a stick and a clock, and you can measure the speed of the light from my torch. And I'll carry a stick exactly the same as yours, and a clock exactly like yours, and I'll measure the speed of light too. First, I want you to be sure that your measuring instruments and mine are identical.'

Vikki put the two sticks side by side – they were exactly the same length. She watched the clocks ticking for a few minutes – they kept identical time. 'Yes, they're the same.'

'Good. Now, as well as measuring the speed of the light from my torch, I want you to keep an eye on my stick. Can do?'

'Sure. And I can chew gum at the same time, too. I'll do my best, Hopper.'

'Right. Let's go for it!'

The Space Hopper hopped past her, and suddenly switched on his torch. Vikki held up her stick and timed the light as it crawled along the stick's length. A quick mental calculation supplied the speed: 300,000 kps. In VUE-foreshortened kilometres, naturally.

She couldn't help noticing that the Space Hopper's stick had shrunk quite a lot, as she'd seen before. It looked about half the length of hers. So he'd see the light travel the length of the stick in half the time she'd measured for *her* stick. So he'd have to measure—

'300,000 kps,' said the Space Hopper.

'No,' protested Vikki, 'you ought to get 600,000 kps. I was watching, and the light travelled along your stick a lot faster than it did along mine. Because your stick had shrunk.'

The Space Hopper shook his horns in mild irritation. 'I've never heard quite so many misconceptions in one statement!' he complained. 'To begin with, you already *know* that you and I are going to measure the *same* speed for light. So why do you think I was going to get 600,000 kps?' The infectious ∪ lit up his face. '2c or not 2c, *that* is the question!' he declaimed theatrically. 'And the answer is . . . not 2c, but c. Then, you say my stick has shrunk. I assure you that from my point of view it hasn't. It's entirely normal. To me, *your* stick has shrunk.'

'Oh. Right. So how come—'

'Think about how we observe each other's *clocks*. According to my clock, light takes the same time to traverse my stick as it does to traverse *your* stick according to *your* clock. We get the same speed, OK? But you observe my stick as being shorter – so what do you deduce about your observations of my clock?'

'Um . . . Oh. Your clock must be slowing down by just the right amount to compensate for the shrinkage of your stick. The light covers a shorter distance, but the clock measures the time as being longer, so the two changes cancel out.'

'As *you* observe my clock and stick, that's a correct description. In other words, from your observational viewpoint, not only do objects contract along the direction in which they are moving, but time also expands by the same amount. If the length halves, the time doubles. That's a consequence of the speed of light being *the same* whether you observe it or I do. Speeds intertwine distance and time: if one changes, so must the other.'

'I guess that makes sense.'

'Yes. Now, it also turns out that an object's mass increases with its velocity, becoming infinite at the speed of light, too – but let's not go into that now. What makes less sense – but is also true in Relativity – is that I don't notice anything strange happening to my stick or my clock – even though you do. Instead, I get the impression that it's *your* stick and clock that are doing funny things. Because, you see, it's just as accurate to say that you're moving relative to me as it is to say that I'm moving relative to you.'

'That's crazy!'

'Planiturthian physics *is* crazy. But that's because the Planiturthian *universe* is crazy. Don't blame the physicists!' The Space Hopper paused. 'Well, not for that, at any rate.' He blinked. 'But you're right, it does seem paradoxical. And, by a remarkable coincidence, the bus is approaching our destination at just the right narrative moment. I want you to meet two acquaintances of mine. The Paradox Twins.'

•

'But – they're *different*!' said Vikki, unable to stop herself. 'You told me they were twins! Oh! I'm sorry! That wasn't very polite of me, was it?'

'Wasn't it?' one of the Space Hopper's acquaintances asked the other.

'No, it wasn't,' said the other.

'But the explanation—' began one.

'Is very—' the other added.

'Simple. We are—'

'—twins.'

'He is—

'—Twindledumb. And I am—'

'No, *I* am—'

'Have it your own way. *He* is Twindledumber.'

What extraordinarily curious people, Vikki thought. 'Being twins would explain why you're the *same*—' she began. 'But not why you're—'

'But we're *not* the same', Twindledumb pointed out.

'No, *he*'s a lot older than—'

'—him.'

'Don't be ridiculous,' said Vikki, 'if you're twins, you must be very nearly the same age as each other.'

'That's the paradox—' the Space Hopper began, but the twins interrupted.

'Do we—'

'—*look* the—'

'—same age?'

'Well, no,' she admitted. 'That's what I was saying right at the start! Twindledumb looks *much* older than Twindledumber.' Then she greyed with embarrassment again. 'Oh! I'm so sorry! I didn't mean to . . . you look *very* distinguish—'

Twindledumb inclined his head of sparse, greying hair to reveal a big bald spot. Twindledumber inclined *his* head of thick, dark hair. 'We're not offended,' they both said at once. 'Yet . . .' they added.

'How . . . how can you be twins, when one of you is much older than the other?' Twindledumb scratched his ear. 'It's a long story—' he began.

'No, it's a short story', contradicted Twindledumber.

'It was to you! To *me*, it was a long one. That's the point!'

Vikki intervened to halt the bickering. 'Surely it has to be either a long story or a short one. It can't be both.'

'Oh, but it can,' said Twindledumb, 'it all depends on who's telling it.'

'No, it all depends on who's *observing* it.'

'Same difference, other way round, yes?'

'No! *Different* difference.'

Once more Vikki tried to calm them down. 'Twindledumb: you said that to you it was a long story. Tell me the story, then – but keep it short. Space Hopper and I haven't got all day, I'm afraid.'

'Tell a long story but keep it short . . . I'll try. About forty years ago, Twindledumber and I were exactly the same age. In fact, we both looked like he does now. But then he went off in a spaceship to a distant star, and he didn't come back until a few weeks ago. That's it. Short enough for you?'

'Admirably concise.'

'It's a pity you don't have time to hear about my onions. I started them from seed, you see, the first year my brother left, and they started to produce *beautiful* little green shoots, and then—'

'Very interesting I'm sure,' said Vikki hurriedly. 'Now, Twindledumber: you said your story was short. So tell it to me. Take your time, no need to rush.'

'Well,' Twindledumber, began, 'it all began when a mutual friend *happened* to remark – it was over dinner, as I recall, munchroom pudding with a salt-cellary glaze – and . . .' He droned on, and everyone except Vikki fell asleep . . . Three hours later, Twindledumber finally reached the point where he climbed into his spaceship and set off for the distant star. 'The ship was fast – *very* fast. Ninety-nine point nine nine nine nine nine nine nine nine nine nine nine nine per cent of the speed of light, the manufacturers claimed, and I'm pretty sure it did at least Ninety-nine point nine nine nine nine nine nine nine nine nine nine nine eight per cent. Anyway, it took me out to the star, we whizzed round the back of it, and its gravitational pull slingshotted us straight back the way we'd come . . . and before I knew it, I was home again.

'To find that *he* . . .' Twindledumber pointed at his brother '. . . had gone prematurely old.'

'Prematurely? *Prematurely?* Forty years I waited for you to get back! Of course I was old!'

'But I was only gone for a couple of days!' The twins glared at each other, and then they both laughed; a little sadly, Vikki felt.

'He's right,' said Twindledumb.

'*He's* right, too,' said Twindledumber.

'We're *both* right,' they said together.

'Relativistic time dilation, Vikki,' said the Space Hopper. 'Because Twindledumber was moving so fast, for *him* time slowed to a crawl. So the whole journey only took just two days. But for Twindledumb, stuck at home and not going anywhere, forty years passed.'

Vikki nodded. Then a thought struck her. 'But . . . just now, when *I* measured *your* clock as running slow, *you* measured *mine* as running slow too! "It's symmetric", you said.'

'I did. And it was.'

'So why isn't *this* symmetric, too? Twindledumb sees Twindledumber's clock as going slower, so to him Twindledumber hardly ages at all, but poor Twindledumb ages forty years. But, by the same token, Twindledumber observes Twindledumb moving very fast relative to *him*, so to him Twindledumb hardly ages at all, but poor Twindledumber ages forty years. Symmetric, see?'

The Space Hopper frowned. 'The difficulty is, Vikki, that after the round trip is over, they're both back in the same place, not moving. They can't *each* be a lot older than the other.'

'Oh. Sorry, never thought of that. But surely that means Relativity is rubbish?'

'Don't be embarrassed. And it's not. What you're missing is a technical point – an important one, as the stories of the Paradox Twins illustrate, but a subtle one. What I said about the speed of light being the same however fast you're moving, but lengths contracting and time expanding – that's true, but only if your speed is uniform. If you *accelerate*, it all gets more complicated. Acceleration is *not* a relative quantity in Alberteinstein's theory. Like I said, "Relativity" is a silly name. "Inertial frames" is the jargon for "what you observe when moving at a constant speed". Even the speed of light ceases to be constant for an accelerating observer. And the source of asymmetry between the twins is that Twindledumber went off in a spaceship, and he accelerated quite fast to get up to lightspeed, and then he accelerated again when rounding the star, and finally he decelerated to avoid smashing into his brother at lightspeed when he arrived home again. Whereas Twindledumb stayed in an inertial frame – no acceleration. So it's *not* symmetric, and that's why Twindledumb has aged forty years while Twindledumber has aged only two days.'

•

Fourday 19 Noctember 2099

Dear Diary

It's not easy getting the hang of Relativity. There are *so* many pitfalls, you need a clear head. Which, frankly, I lack at the moment.

For instance, there's the paradox of the near-lightspeed train. Suppose that a train is being driven along a railway track at close to lightspeed, and there are small gaps between lengths of the track. Then the driver observes the gaps as moving at near-lightspeed, so they *shrink*, and to him they appear even smaller than they would have been when he was (they were?) stationary. So he expects just the tiniest of bumps as the train passes over each gap.

On the other hand, to a 'stationary' observer beside a gap, the TRAIN is moving at near-lightspeed, so *it* shrinks – so much so that it ought to fall into the gap between the rails!

How do you resolve *that* contradiction, Diary?

You don't know, do you? The answer, as in all of these apparent paradoxes of Relativity, is to remember that relativistic geometry is mathematically self-consistent. Any apparent problems must result from its interpretation. So you use the maths to work out what really happens. In this case, the train *does* 'fall into' the gap – but it's travelling so *fast* that it doesn't fall very *far*! So, even as viewed by a stationary observer beside the gap, all that happens is a very tiny bump.

Which is what the driver experiences, too.

So, if you want to think sensibly about relativity, Diary Dear, then you have to go the whole hogsagon. A mixture of nonrelativistic and relativistic thinking is inevitably fatal.

13

THE DOMAIN OF THE HAWK KING

'They were *such* a sad pair,' said Vikki. 'I'd hate to wait forty years for my brother to return from a voyage, only to discover he'd hardly aged at all compared with me.'

'It's even sadder. You see, Twindledumb could have gone to the star with his brother, but he decided not to.'

'Why?'

'Said he couldn't spare the time.'

There was a long silence.

'Was that . . . a joke?' Vikki asked, finally.

'I wish it had been. It shows how important it can be to understand your universe as it is, and not just as you imagine it to be.'

'I suppose. Still, Twindledumb could always make an interstellar voyage himself, and get their ages back in synch.'

'I'm not sure the spaceship still exists. But there's another way that Relativity could help the twins overcome their paradoxical temporal displacement.'

Vikki waited for him to continue, but he didn't. So eventually she asked, 'What?'

'What what?'

'What *way*. To help the twins.'

'Oh, sorry, my mind was wandering. Time travel.'

'You mean – going into the future or the past?'

'We *all* travel into the future – at one year per year. The past . . . well, that's trickier. But perhaps not impossible. Relativity doesn't just change our perceptions of space, time, and matter. It also affects our view of *causality*.'

'Which is?'

'The link between cause and effect. If you want to understand that, we need to visit Minkowski Space. Now, because of the commercial exploitation of Intellectual Property rights . . . look, it's tricky to explain, but basically somebody *bought* Minkowski Space, along with several other spaces that you really ought to visit. They've been franchised out for the tourist trade, so I'll need to pull a few strings to get us in there without paying. I'll just make a call on my mobile.'

The conversation was lengthy, laced with words that Vikki didn't understand, and mostly consisted of the Space Hopper reminding whoever was on the other end of past favours granted and future favours to come. Occasionally the Space Hopper flashed her his trademark ∪, as if to indicate that the negotiations were succeeding, and once he actually *winked* at her.

'OK, we're in – but only if we can get there in a hurry. Let me reset your VUE, and we can head straight for—'

•

'—Minny Space. And there she *is*! Hi, babe! Long time no see!'

'Don't you try your alien charms on me, Hopper. If you wanna be my Hopper, *you* have got to *give*, baby. Who's the geom?'

'What's a geom?' asked Vikki.

'Slang for "geometric entity",' said the Space Hopper. 'As a Flatland line, you're an archetypal geom. Don't be offended – it's just Minny's way. She does it to maintain street cred.'

Vikki decided not to ask what street cred was. Life was too short to sort out *everything* you didn't understand. 'Minny?'

'Minny Space. Short for Minkowski Space, you see . . . She's one of the Space Girls. And if I'm not mistaken, here come the others! Victoria: meet Curvy Space, Bendy Space, Pushy Space, and Squarey Space.'

Suddenly Vikki was surrounded by strangely clad – well, that wasn't exactly the word, but it would have to do – Spaces. Rather brash, fast-talking spaces, who used a lot of street slang and for some reason kept breaking into song-and-dance routines. They seemed very confident, very self-aware, and – to be frank – they were a considerable pain in the endpoint.

When they discovered that Vikki was a girl, though, they immediately tried to make her feel at home. And, apparently, to empower her – an operation for which they were an endless fount of novel ideas. 'Yo, Vikki, what you Flatty girls need is *Line Power*!'

'I've heard of a power line,' said Vikki. 'Would that do?'

'No, that's *quite* different!' pouted Pushy Space.

'The Lines of Flatland should assert their power!' asserted Squarey Space.

'Stop giving in to the Polygons!' Curvy Space added. 'Right now! Thank you very much!'

'Above all, get yourself a Good Manager!' advised Bendy Space.

'Exploit your assets!' Minny Space clarified.

'Kick some assets!' shouted Curvy Space, and the others all laughed.

This all sounded like fun, and Vikki would have tried it if she'd had the foggiest idea what any of them were talking about. Female emancipation, of some kind, by the sound of it. 'We're working on it', she told them. 'At least, *I* am. I don't think my mum is. But I kind of suspect she secretly approves. It takes a while . . . I'm not sure I really know—'

'What part of "know" don't you understand, girly?'

'Minny,' the Space Hopper interjected, 'could you oblige and introduce Vikki to modern notions of causality?'

'Yeah, sure. OK, Vikki – pay attention and ignore the grinning idiot who brought you here. In relativistic spacetime, everything depends heavily upon which "frame of reference" an observer uses. Moving and static observers see the same events in different ways. Now, in full-blooded Minkowski spacetime, space is 3D; but I'm Minny Space, and that's not just a nickname like the Space Hopper told you, it's because I'm a miniature version. My spatial coordinate is one-dimensional.

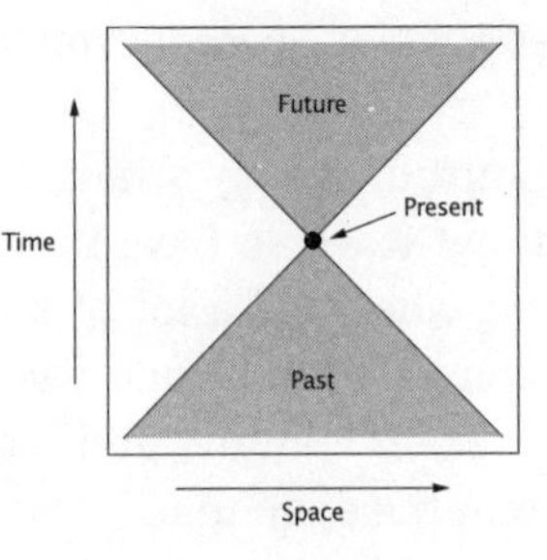

'Now, as a particle moves through spacetime, it traces out a curve, its *world-line.* If its velocity is constant, its world-line is straight. The slope of the world-line depends on the particle's speed. Particles that move very slowly cover a small amount of space in a lot of time, so their world-lines are close to the vertical; particles that move very fast cover a lot of space in very little time, so their world-lines are nearly horizontal. Got it?'

'Sure have.'

'In between, at an angle of 45 degrees, are the world-lines of particles that cover a given amount of space in the same amount of time – measured in the right units. Those units are chosen to correspond with each other through the medium of the speed of light – say years for time units and light-years for space units. Tell me, Vikki, what covers one light-year of space in one year of time?'

Vikki knew that one. 'Light.'

'Right on, babe, and I choose my words with care. So world-lines sloping at 45 degrees correspond to light rays, or anything else that can move at the same speed as light.'

'My head is starting to ache.'

'You ain't seen nothin' yet! Now, Relativity won't let anything move faster than light. There's a maths reason: if things could move faster than light their lengths would become imaginary, and so would their masses and the local passage of time. So the world-line of a real particle can never slope more than 45 degrees away from the vertical. Such a world-line is called a *timelike curve.* The extreme opposite is a curve in Minny Space that *always* slopes more than 45 degrees away from the vertical. That's called a *spacelike curve,* and no part of it can coincide with the world-line of a particle.

'Any event in space-time has associated with it a *light cone,* formed by the two 45 degree diagonal lines that pass through it. It's called a cone because when space has two dimensions, the corresponding surface really is a (double) cone. The forward region contains the *future* of the event – all the points in spacetime that it could possibly influence. The backward region is its *past* – all the events that could possibly have influenced *it.* Everything else is forbidden territory, elsewheres and elsewhens that have no possible causal connections with the chosen event.

'A timelike curve always moves from its past into its future. Spacelike curves are quite different: no point on them lies in the past or future of any other point.

'In ordinary space, the distance between two points tells you how far apart they are. In Special Relativity there's something kinda similar, and it's called the *interval* between spacetime events. Along the 45 degree lines the interval is zero, so those lines are called *null curves.* The interval tells you how time seems to pass for a moving observer.'

'Yes, the Space Hopper told me about that,' said Vikki. 'It's time dilation, isn't it? The faster an object moves, the slower time appears to pass for an observer in a frame that moves with it.'

'Yes. If you could travel *at* the speed of light, time would be frozen. No time passes on a photon.'

'So light is frozen time?'

'I suppose you could put it like that.'

•

The Space Hopper was rabbiting on about time travel again.

'But I thought there were paradoxes associated with time travel,' said Vikki. 'For instance, suppose I could go back and warn ancestor Albert that his messing about with the Third Dimension would land him in the pokey! Then he might never have written *Flatland* . . . but then I wouldn't have known about him, so I *wouldn't* have gone back and warned him . . . so he would have written the book and *then* – well, you see what I'm getting at!'

'I certainly do,' said the Space Hopper. 'A concise statement of the Great-great-grandfather Paradox. But we Space Hoppers generally find that it's best to find out what the universe is capable of by trying it out, rather than *ruling* it out because of philosophical paradoxes. The paradoxes have a habit of taking care of themselves, you know. Anyway, we generally find we have more fun that way. Before I can tell you about time travel, though, you have to understand how gravity fits into a relativistic picture of the universe.'

'What does gravity have to do with time travel?'

'Everything. Though I admit it's not obvious. You see, Alberteinstein invented another theory, called General Relativity, which was a combination of Newtonian gravitation and Special Relativity. You know what Isaacnewton said about gravity?'

'No, I've not studied Planiturthian history in that much detail, Hopper!'

'He said it was a force that moves particles away from the perfect straight lines they would otherwise follow. He worked out a law for how the gravitational force exerted by any particle of matter varies with distance.'

'OK, I remember that.'

'Alberteinstein wasn't keen on forces. He preferred to think *geometrically*. The paths that particles follow, in the absence of any forces such as gravity, are called *geodesics*. They are shortest paths – they minimize the total distance between their endpoints. In flat Minkowski spacetime – Minny Space and her full-blown extension to 3D space plus 1D time – the analogous relativistic paths minimize the interval instead.' Vikki had difficulty imagining anything more full blown than Minny Space. 'The problem is to add in effects of gravity consistently. Alberteinstein incorporated gravity not as an extra force, but as a distortion of the structure of spacetime, which changes the value of the interval. This variable interval between nearby events is called the *metric* of spacetime. The usual image is to say that spacetime becomes "curved".'

'Curved round what?'

'That's why Curvy Space is here. She can show you.'

•

'It's not curved round anything, girly. It's just intrinsically distorted compared with flat spacetime. You might as well ask "flat along what?" about ordinary Euclidean space, it's just as sensible – or silly – a question. The curvature is interpreted physically as the force of gravity, and it causes light cones to deform.'

'How can space be curved?' asked Vikki.

'I'm surprised a Flatlander who's seen the Mathiverse can ask that,' said Curvy Space. 'You used to think space had to be flat. Now you know better!'

'Yes, but that was 2D space, and it's easy to imagine a curved surface. But a curved solid?'

'Platterland was curved,' said the Space Hopper, 'but when we were on it, it didn't look *bent*.'

'You mean the disc had a curved edge?'

'No, not the edge – the whole *space* was curved. That's what made the geometry so weird. Platterland has constant negative curvature. A sphere has constant positive curvature. And a plane, like Flatland, has zero curvature. Different curvatures, different geometries. You can get curvature in 3D, too.'

'How?'

'You're suffering from too heavy a Spaceland diet, girly,' said Curvy Space. 'Tell me, what would a curved 2D universe look like *from inside*?'

'Bent?'

'Not at all! Flat. Because light would follow the surface, see? It would go along the geodesics – the shortest paths *that lie in the surface*.'

'Then how could you tell it wasn't flat?'

'Because images would distort as light rounded the curves', said the Space Hopper. 'Distances wouldn't match up to the model of a Euclidean plane. Which is why the shortest paths, the geodesics, in Hyperbolica looked – from an outside VUEpoint – like circles.' She still looked uncertain, so he cast around for another example. 'Remember the squarrel? As soon as you see a five-sided squarrel with all angles 90 degrees, you *know* you're in a negatively curved space. And the same goes for curved 3D space, or curved 4D space-time.'

'Yes, but . . . what would the space be curved *around*?'

The Space Hopper grunted, and Vikki suddenly found herself falling into Squarey Space.

•

'Don't ask "curved around what", baby. Ask "*flat* around what"!'

'That makes no sense at all.'

'Yes it does, kid. Look at me – I'm flat.'

'You *look* flat. But Curvy Space just told me that doesn't mean anything.'

'Believe me, take my measurements, I'm as flat as a pancake in a steamroller rodeo. But I'm a torus.'

'Well, I'm Libra, but I don't actually believe in that sort of—'

'No, no. A topological doughnut. And just watch it, OK?'

'I wasn't going to say a word about what kind of nut you—'

'I said, watch it! Know why they call me Squarey Space?'

'No.'

'Because ultimately I'm a flat square.'

'But you said you're a torus, and that's all curves.'

'Ah. A torus embedded in 3D is curved. An unembedded one, just being its own space without any outside assistance, can be flat.'

'How?'

'Take a look over there – no, a bit further to your left. What do you see?'

'Er . . . wow! It looks like *me*!'

'It is you. And beyond that?'

'An even tinier me. And another, and another . . . fading into the distance. The light rays are bending round like a projective lion.'

'The light rays are bending *straight* like a projective lion, babe. In certain directions, you see yourself. That's because opposite edges of my underlying square *wrap round* and glue together, seamlessly. You don't need to bend me to make them join – you just have to declare that they do. Or, if you prefer, you can tile an infinite plane with copies of that square, and insist that whatever happens in one copy, happens in all of them. It's the same thing.

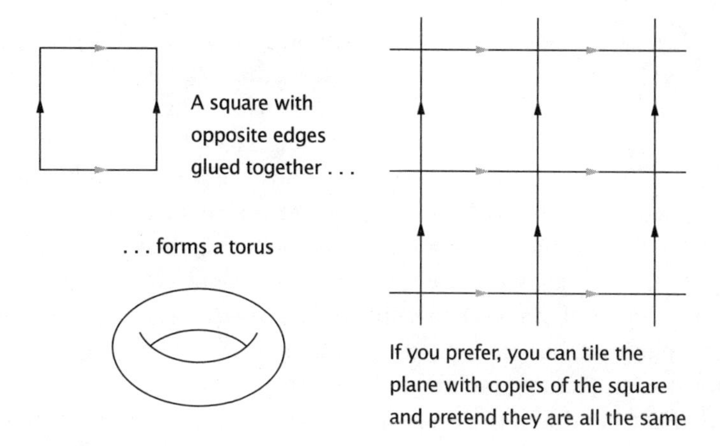

'So, you see, I'm flat, and yet I have closed geodesics. Which don't bend *around* anything, because there *isn't* anything else anyway. Intrinsically, a space can be bounded, have no edges, and yet be flat. Compared to that, curved 3D space is a piece of cake. Just squeeze its metric here and there.'

•

With a PLOP! Vikki found herself back in metaspace. 'So now you know,' said Squarey Space.

'And when the Planiturthians made *very* careful measurements,' Pushy Space butted in, 'they discovered that their universe is curved.'

'How did they discover that?'

'Look for the rainbow in every storm, kid. Bent light, you dig? Curved space leads to "gravitational lensing", where light gets bent by heavy objects. You can see it happening during an eclipse of the Sun. And quasars – powerful, distant kinds of superstar – produce multiple images in telescopes because their light is lensed by a galaxy that is in the way.'

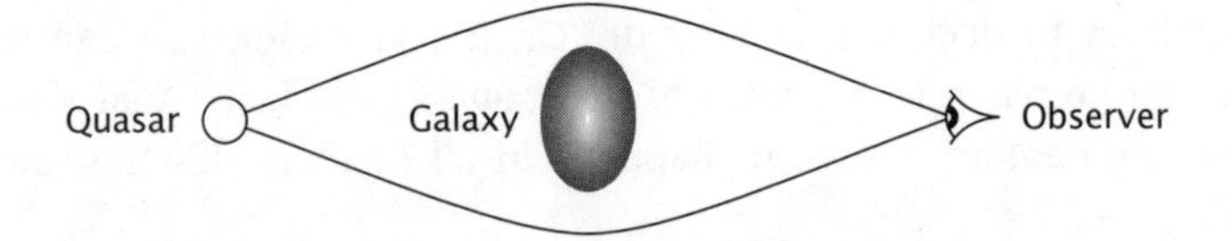

'It's easier to visualize,' the Space Hopper said helpfully, 'if you think of a spacelike section of 3D spacetime – 2D space, 1D time – near a star. It forms a curved surface that bends downwards to create a valley, and the star sits inside. Light follows geodesics across the surface and gets "pulled down" into the hole because that path provides a short cut. Just like the circular arcs – to our eyes – that form geodesics in Hyperbolica. Particles moving in spacetime at sub-lightspeed behave in the same way. If you look down from above you see that the particles no longer follow straight lines, but are "pulled towards" the star, which is where the Newtonian picture of a gravitational force comes from.

'Far from the star,' the Space Hopper continued, 'this spacetime is very close to Minkowski spacetime – the gravitational effect falls off rapidly and soon becomes negligible. Spacetimes that look like Minkowski spacetime at large distances are said to be *asymptotically flat*. Remember that term: it's important for making time machines.' He was back to those again. 'Most of our own universe is asymptotically flat, because massive bodies such as stars are scattered very thinly.'

Vikki digested this information. 'So we could give spacetime any form we wanted? That sounds a bit *too* flexible.'

'No. When you're setting up a spacetime, you can't just bend things any way you like. The metric must obey the *Einstein equations*, which relate the motion of freely moving particles to the degree of distortion away from "flat" Minkowski spacetime.'

'You mean,' said Vikki, 'that there's a connection between the distribution of masses within a spacetime and the structure of the spacetime itself? As if matter . . . creates and moulds its own space and time?'

'Exactly.'

•

Fiveday 20 Noctember 2099

OK, Diary: finally, the Space Hopper explained how to interpret the phrase 'time machine' within the framework of General Relativity.

Here's the gen. A time machine lets a particle or object return to its own past, so its world-line, a timelike curve, must close into a loop. A time machine is just a *closed timelike curve*. So instead of asking, 'Is time travel possible?' we ask, 'Can closed timelike curves exist?'.

So, can they?

Not in Minny Space. There, forward and backward light cones – the future and past of an event – never intersect. But they can intersect in other types of spacetime. The simplest is Minny Space rolled up into a cylinder. Then the time coordinate becomes cyclic, but the spacetime is still flat.

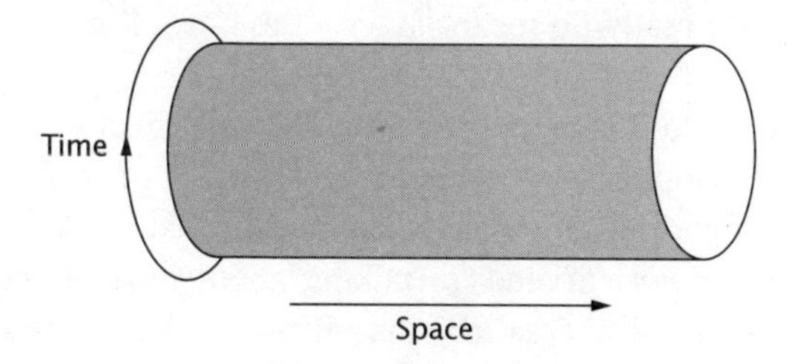

In such a spacetime, history repeats itself, over and over again. Sort of, anyway. *Spacetime* repeats – but what happens to history depends upon whether you think free will might be in operation. It's a tricky question, and one that Einstein's equations don't really address. They just govern the overall coarse structure of spacetime.

I told you that this cylindrical spacetime is flat, Diary Dearest, though I admit it doesn't look that way. It's rather like Squarey Space, but with time thrown in, so you have to be careful. Although a cylindrical spacetime *looks* curved, actually the corresponding spacetime is *not* curved – not in the gravitational sense. When you roll up a sheet of paper into a cylinder, it doesn't *distort*. You can flatten it out again and the paper isn't folded or wrinkled. A creature that was confined purely to the surface wouldn't notice that it had been bent, because distances *on* the surface wouldn't have changed. In short, the metric – a local property of spacetime structure *near* a given event – doesn't change. What changes is the global geometry of spacetime, its overall topology.

•

'Rolling up Minny Space,' said the Space Hopper, 'is an example of a powerful topological trick for building new Spacetimes out of old ones: *cut-and-paste*. If you can cut pieces out of known Spacetimes, and glue them together without distorting their metrics, then the result is also a possible Spacetime.'

'You're speaking metaphorically, of course.'

'Well, until recently I'd have agreed with you. But now the Hawk King's moved into the spacetime construction business.'

'The Hawk King?'

'A very, *very* impressive being. I'm planning an audience with him, if we can swing it. When it comes to spacetime heavy engineering – and I really do mean *heavy* – the Hawk King is in a class of his own. But I'm getting ahead of myself.'

'Well, you said you were interested in time travel.'

The Space Hopper laughed politely. 'I say "distorting the metric" rather than "bending", for exactly the reason that I claim rolled-up Minny Space is *not* curved. I'm talking about intrinsic curvature, as experienced by a creature that lives in the spacetime, not about apparent curvature as seen in some external representation. Apparent bending of this type is harmless – it doesn't actually change the metric. Now, the rolled-up version of Minny Space is a very simple way to prove that Spacetimes that obey the Einstein equations *can* possess closed timelike curves – and thus that time travel is not inconsistent with currently known physics. But that doesn't imply that time travel is *possible.*'

'I see that. There's a distinction between what's mathematically possible and what's physically feasible.'

'Very good, Vikki. A spacetime is mathematically possible if it obeys the Einstein equations. It's physically feasible if it can exist, or could be created, as part of a specified universe. Which is where the heavy engineering comes in. Unfortunately for time travel in the Planiturthian universe, there's no reason to suppose that rolled-up Minkowski spacetime is physically feasible: certainly it would be hard to refashion the Planiturthian universe in that form if it didn't already have cyclic time. The search for Spacetimes that have closed timelike curves *and* have plausible physics is a search for more plausible topologies. There are many mathematically possible topologies, but you can't get to all of them from a given starting point. But you can get to some remarkably interesting ones, based on black holes.'

'Are those like topological holes?'

'Only very loosely speaking, Vikki. You know that in classical Newtonian mechanics there's no limit to the speed of a moving object, so particles can escape from an attracting mass, however strong its gravitational field, by moving faster than the appropriate escape velocity. In 1783 the Planiturthian astronomer Johnmichell realized that when this idea is combined with a finite velocity for light, it implies that really heavy objects can't emit light at all – because the speed of light is lower than the escape velocity. The light travels too slowly to be able to get away, you understand?'

'Sure. That's easy.'

'In 1796 another astronomer, Pierresimondelaplace, put the same idea in a book of his called *Exposition of the System of the World.* And both Planiturthian astronomers imagined that the universe might be littered with huge bodies, bigger than stars, but totally dark.'

'Wild!'

'You said it. They were both a century ahead of their time. In 1915 Karlschwarzschild asked whether the same kind of thing could happen with relativistic gravitation. He solved the Einstein equations for the gravitational field around a heavy sphere in a vacuum. His solution behaved very strangely at a critical distance from the centre of the sphere, now called the *Schwarzschild radius.* When it was first discovered, People thought this strange behaviour just meant that space and time lost their identity in Schwarzschild's solution, and became meaningless. A mathematical artefact, yes? With no physical implications at all. But then some bright spark asked what would happen to a star so dense that it lies completely inside its own Schwarzschild radius.'

'And what *would* happen, Hopper?'

'It would collapse under its own gravitational attraction! In fact, a whole portion of spacetime around that star would collapse to form a region from which no matter, not even light, could escape. In 1967 Johnarchibaldwheeler called such a region a *black hole*, and the name stuck.'

'Does a black hole stay the same for ever, or does it change?'

'It changes. First the star shrinks symmetrically until it hits the Schwarzschild radius, after which it continues to shrink more rapidly until, after a finite time, all the mass has collapsed to a single point, the *singularity*. From outside, all you can detect is the *event horizon* at the Schwarzschild radius, which separates the region from which light can escape from the region that's forever unobservable from outside. Inside the event horizon lurks the black hole.

'If you were to watch the collapse from outside, you'd see the star start to shrink towards the Schwarzschild radius, but you'd never see it get there. As it shrinks, its speed of collapse as seen from outside approaches the speed of light, and relativistic time dilation means that the entire collapse takes infinitely long when seen by an outside observer. However, the collapse time experienced by an observer on the surface of the star would be finite. Once inside a black hole, the roles of space and time are reversed. Just as time inexorably increases in the outside world, so space inexorably decreases inside a black hole.

'That's where the scope for engineering comes in,' said the Space Hopper. 'The Hawk King has developed a whole battery of techniques, from quantum foam enlargement to improbability calculus. Now, the spacetime topology of a black hole is asymptotically flat – its "mouth" opens out and becomes flat at large distances. So a black hole can be cut-and-pasted into the spacetime of any universe that has reasonably large asymptotically flat regions – such as the Planiturthian one. This makes black hole topology physically plausible in such a universe. The scenario of gravitational collapse makes it even more plausible: if you want to build a black hole, you just have to start with a big enough concentration of matter, like a neutron star or the centre of a galaxy. That's what I meant by heavy engineering. The technology of 3001 will be able to *build* black holes. With matter processors – modified neutron stars with gravitational traps and heavy-duty laser-compressors.

'However, a black hole isn't enough, because a static black hole doesn't have closed timelike curves. However, Einstein's equations are time-reversible: to every solution of the equations there corresponds another that is just the same, except that time runs backwards. The time-reversal of a black hole is a *white hole.* An ordinary event horizon is a barrier from which no particle can escape. A time-reversed event horizon is one into which no particle can fall, but from which particles may from time to time be emitted. So, seen from the outside, a white hole would appear as the sudden explosion of a star's-worth of matter, coming from a time-reversed event horizon.'

'Why should the singularity inside a white hole suddenly decide to spew forth a star, having remained unchanged since the dawn of time?' protested Vikki.

'Good point. It makes causal sense for an initial concentration of matter to collapse, if it's dense enough, thereby forming a black hole; but the reverse seems to violate causality. It doesn't, of course – but the cause would have to lie outside our own universe, beyond the white hole's event horizon, so we wouldn't see it coming. Let's just agree that white holes are a mathematical possibility, and notice that they're also asymptotically flat when they open out enough. So if you knew how to make one, you could glue it neatly into your own universe. The Hawk King has just developed an effective method for doing that, based on the uncertainty principle. He uses a Heisenberg amplifier to make the position of matter so uncertain that it may well be outside the normal universe altogether. Not only that, he can also glue a black hole and a white hole together. He cuts them along their event horizons with a cosmotome and sews the edges together with exotic – that is, negative-energy – matter.'

'And what does that give?'

'A wormhole. A sort of tube. Matter can pass through the tube in one direction only: into the black hole and out of the white hole. It's a kind of matter valve. And the passage through the valve is achieved by following a timelike curve, because material particles really can traverse it.

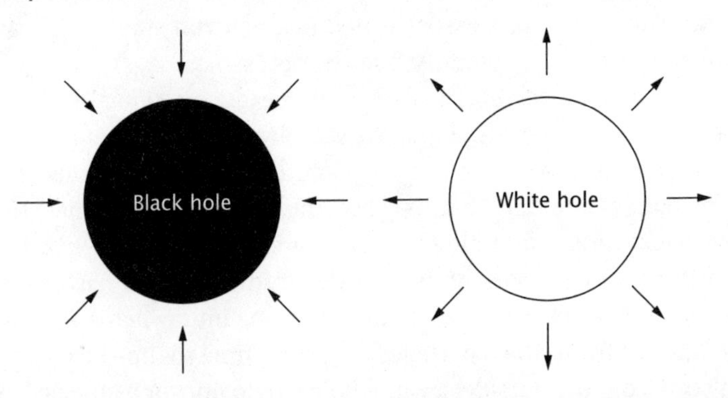

'Now comes the clever bit. Because the topology of spacetime is asymptotically flat at both ends of the tube, both ends can be glued into any asymptotically flat region of any spacetime. You could glue one end into the Planiturthian universe, and the other end into somebody else's; or you could glue both ends into the Planiturthian universe – *anywhere you like*, except near a concentration of matter. Now you've got a – guess.'

'A spacetube?'

'No: a *wormhole.* The Hawk King makes the best wormholes in the universe. They're called wormholes because they look like the holes a maggot bores in an apple. Only here the apple is – well, not so much spacetime as everything that's *not* spacetime.

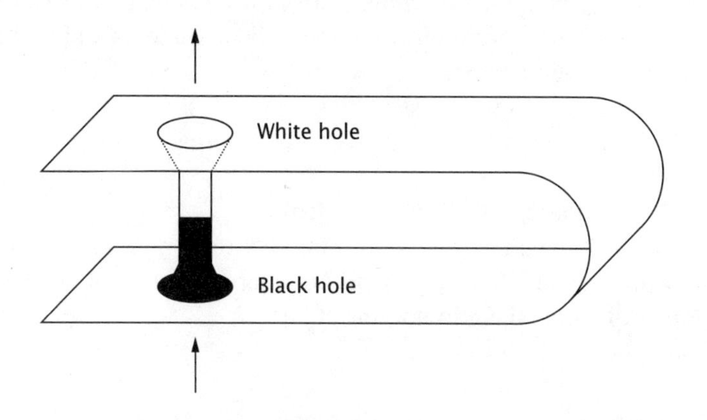

'The next important thing is this: the distance *through* the wormhole is very short, though the distance between the two openings, across normal spacetime, can be as big as you like.'

'I see,' said Vikki. 'A wormhole is a short cut through the universe.'

'Right,' said the Space Hopper.

'But that's *matter transmission,* not time travel.'

'So far, yes,' said the Space Hopper. 'But there's more.'

•

'It ought to be easy to use the VUE to explore a wormhole,' said the Space Hopper. 'Unfortunately, it's not. Most parts of the Mathiverse are in the Public Domain, because mathematicians never understand how important their ideas are until long after they've told everyone about them. Anyway, according to metaspatial law, mathematicians aren't allowed to patent mathematics because it consists of *ideas.* Anyone *else* can patent mathematics, as long as they don't admit that that's what it is, but mathematicians can't because – being mathematicians – they *know* it's mathematics, and are bound by their logical training to say so.

'The Hawk King has acquired a vast commercial empire (so I

suppose he ought to be called the Hawk Emperor, but somehow that doesn't sound right) by securing a monopoly on Time Travel mathematics. In principle he also controls the Mathiverse's supply of magnetic monopoles, so he has a Monopole Monopoly – but since no one has found a monopole yet, that's a bit hypothetical. Since all rights to time machinery are controlled by the Hawk King, we will have to request an audience with His Majesty and petition him for temporal access.'

'How do we do that?' asked Vikki.

'We VUEtravel to where he lives, and approach the appropriate officials.'

'Where does His Majesty live, then?'

'In the Domain of the Hawk King. It's right next to the Public Domain, separated by a domain boundary, so a couple of Space Hops will get us there in no time at all . . .'

•

'Passport? What do you mean, passport?'

'No one enters the Domain of the Hawk King without a passport,' said the official from His Majesty's Customer Extortionists.

'But we haven't got—' Vikki began.

'What *sort* of passport?' asked the Space Hopper. 'Would I be right in thinking that it's made of paper?'

'Of course.'

'Strong paper?'

'Indeed.'

'With lots of fancy engraving on it?'

'Definitely on the right track.'

'And it sort of . . . *folds*?'

'Yup, sure sounds like a passport to me,' said the official.

'Would that be a big piece of paper, or a small one?'

'Oh . . . big. Yes, definitely. Big.'

'How big?'

'How big have you got?'

'How big will secure us an immediate audience with the Hawk King?'

'Ooooooh . . . about five times as big as your run-of-the-mill passport.'

'Would *this* be suitable?' There was a rustling as paper changed hands.

'Definitely. Your documents are completely in order, Mr . . . er . . . Bankofspaceland. And your good lady Ms Travellerscheque too, I'm pleased to say. Let me just call up His Majesty's schedule . . . Right. You have a five-minute audience at 3.50 this afternoon. Just give your ticket to the official at the door.'

'But we don't have a—' Vikki began.

'Vikki, the ticket is without doubt a document *just like a passport.*'

'Uh – oh, right, Space Hopper. I see.'

•

The Hawk King's audience room was vast. He sat on a splendid throne at the far end. It took them *for ever* to make their way towards him.

'This is where a time machine would really come in useful,' the Space Hopper whispered to Vikki.

'Yes. Or a wormhole. I could do with a short cut.'

'I'd be careful using that sort of metaphor in the Hawk King's presence,' cautioned the Space Hopper. 'He has a habit of taking things literally.'

'Ulp.'

Eventually they arrived at the foot of the throne. The Hawk King's piercing eyes dissected their souls.

'You do realize,' he said, 'that people used to think time travel was a theoretical impossibility, a contradiction in terms?'

Obviously he had overheard their whispered conversation. Hawks reputedly have astonishing eyesight – perhaps his hearing was just as acute. Or perhaps the audience room was equipped with sensitive microphones.

'Of course, now we know better,' the monarch continued. 'You are thinking of exploring one of my wormholes?'

'Uh – my companion did *mention* a wormhole, Your Majesty.'

'You cannot afford the price, you realize.'

'We were hoping that wouldn't be necessary, Your Majesty,' said the Space Hopper.

'Is there a connection between wormholes and time machines, Your Majesty?' Vikki asked in a squeaky voice.

'Of course. It was the Paradox Twins that suggested the idea to me originally – you've met them? Yes, I see you have. Never mind, you'll get over it. They experienced a time *discrepancy* – but it led into the future, not the past. However, in conjunction with a worm hole – that time discrepancy can be turned into a closed timelike curve.'

'Uh – how, Your Majesty?'

'Fix the white end of the wormhole, and tow the black one away – or better still, zigzag it back and forth – as close to lightspeed as you can get.

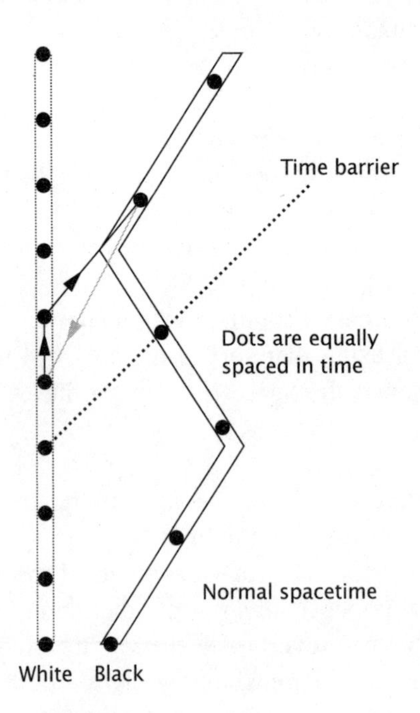

'The white end of the wormhole remains static, and time there passes at its normal rate. The black end zigzags to and fro at just less than the speed of light. So time dilation comes into play, and time passes more slowly for an observer moving with that end.'

'Ah!' said the Space Hopper. 'I see, Your Majesty. A brilliant insight!'

'Um—' Vikki began.

'Why, Vikki – all you have to do is think about world-lines that join the two wormholes through normal space, so that the times experienced by observers at each end are the same', said the Space Hopper. 'At first, those lines have a slope of less than 45 degrees, so they're not timelike, and it's impossible for material particles to travel along them. But as the black end of the wormhole zigzags back and forth, at some moment that line achieves a 45-degree slope. And once this "time barrier" is crossed, you can travel from the white end of the wormhole to the black one *through normal space* – following a time-like curve. And once you've arrived there, you can return *through* the wormhole, again along a timelike curve. Because the wormhole forms a short cut you can do that in a very short time, effectively travelling instantly from the black end to the white end. And now you're in the same place as your starting point, but in the *past*!'

'Indeed,' said the Hawk King. 'You have travelled back in time. And the actual distance you travel through ordinary space can be quite small – it depends on how far the black end of the worm-hole moves on each leg of its zigzag path. In space of more than one dimension it can spiral rather than zigzag, which corresponds to making the black end follow a circular orbit at close to light-speed.'

'Wonderful, Your Majesty,' said the Space Hopper. 'Vikki, you could do that by creating a binary pair of black holes, rotating rapidly round a common centre of gravity!'

Vikki gave the matter thorough consideration. *Somebody* might be able to do that. Not her. 'Would I be right in thinking that the longer you wait before you start, the further back in time you can go?'

'Yes,' said the Space Hopper, 'but there's a nasty snag. You can never travel back past the time barrier, and that occurs some time *after* you build the wormholes! So there's no hope of going back to the time of the Planiturthian dinosaurs, or the time of the Universal Colour Bill in Flatland.'

'Unless someone found a very old, naturally occurring worm-hole,' the Hawk King pointed out.

'Is that what you've done, Your Majesty?'

'That, my dear young Line, is a commercial secret. And the time allocated for your audience has almost run out, so I strongly suggest that you present your petition immediately.'

Royal hints should not go unheeded. 'We should like to experience time travel', said the Space Hopper.

'Into the past,' Vikki added quickly, just in case there was a misunderstanding.

'Ah. And do you have a reason for this wish?'

Vikki racked her brains . . . why *did* they want to try time travel? 'To see if we can help the Paradox Twins, Your Majesty.'

The Hawk King rose and spread his magnificent feathers. Vikki and the Space Hopper tried hard not to cower. 'That is a worthy motive,' the King remarked. 'I have decided to grant your request. My aides will deal with the details on your way out. You are dismissed.'

14

DOWN THE WORMHOLE

Sevenday 22 Noctember 2099

Thus, Dear Diary, began the most curious adventure that I have yet experienced during my Misguided Tour of the Mathiverse! The King's aides took us into a special room, full of strange, arcane machinery. And in the middle of that room, hovering unsupported, was—

•

'—a *red* hole?' said Vikki, disappointed. It was like a big blood-red bubble, and it *sparkled*. 'I thought they were black.'

'A common misconception – or perhaps I should say "misperception",' said the Master of the Royal Wormhole. Vikki had taken an instant dislike to him: the Hawk King was majestic and awe-inspiring, whereas the Master of the Royal Wormhole was an arrogant snob. She wasn't sure they ought to trust him. But the Space Hopper seemed unconcerned. 'The term "black hole" refers to the manner in which light is sucked into the singularity. However, thanks to relativistic time dilation, the light emitted by anything that is falling into a black hole actually appears red.'

'Oh.'

'For example, the bunch of roses that you see sinking slowly towards the event horizon – the surface of what appears to be a bubble – actually crossed the event horizon several weeks ago, and by now has emerged into a previous century. What we see is the light that the roses emitted, which has spent the intervening time struggling painfully away from the black hole's clutches. And struggling light shifts towards the red end of the spectrum. Not only that, but every black hole is surrounded by a complete record of everything that ever fell into it. But unless the object fell in recently, the image is extremely squashed and redshifted, and mixed up with all the images of all the other objects. And the longer ago an object fell in, the closer it had to get to the event horizon for its image still to be visible. Still, it's an intriguing thought, is it not?'

'So . . . er . . . how do we *use* the wormhole?' asked the Space Hopper.

'Easy. You fall into it.'

'How do we do—'

'The problem, young lady, is to *stop* yourself falling into it. You see that white line marked on the floor?'

'Yes.'

'Step a micrometre beyond that and the Black Hole will suck you in. Guaranteed.'

'Oh.'

'But the gravity will shred you like cheese in a grater.'

'Space Hopper, I think I've changed my mind—'

'Unless you wear protective suits of exotic matter.'

'Ah. That's different.'

'Which are available for a small – um – *documentary* exchange.'

More rustling. Weird suits that didn't quite seem to be *present* were brought, and the two travellers climbed into them. After a perfunctory safety check they were led towards the fateful white line.

The tip of one of the Space Hopper's horns crossed the line. Suddenly his shape was drawn out into a long, thin tube. His body elongated, then flattened – then shrunk as if all his insides had been sucked away along an unseen dimension. He became a dwindling speck of redshifted light.

Vikki took a deep breath, and followed him in.

•

Wunday 23 Nocternber 2099

Diary: it was the second most *HORRIBLE* experience of my LIFE!!!!

What was the FIRST most horrible?

I'm coming to that.

•

'We're trapped!' said the Space Hopper.

They were circling the black hole's singularity, held away by the forces generated by the suits' exotic matter. They should have been swept past it towards the white hole exit from the wormhole.

But there was no exit.

'It was just an ordinary black hole all along,' complained the Space Hopper. 'There never was a way out. The Master of the Royal Wormhole was playing games with us. Very nasty games. I imagine he pocketed the cost of the white hole construction instead of getting it built. And now we're stuck. I'm sorry – I should have realized he wasn't to be trusted.'

'I didn't like him,' said Vikki, 'and I should have said so. But can't we use the VUE to get out?'

'It doesn't work in here. Exotic matter interferes with its conceptical actuvailers.'

'Then we're *trapped*!'

'I already said that.'

'I know – and it's still true! Is there a way out?'

'Does it look like there's a way out?'

'No, but there *must* be! We could *die* in here!'

The Space Hopper gave her a mournful look. 'I'm afraid that's exactly what we will do', he said. 'No food, no drink – and no way out. And I don't think there's any kind of "and with one bound our heroine was free" trick that can save us.'

There was a long silence.

'We're stuck, then,' said Vikki disconsolately.

'You certainly are,' said Vikki.

Vikki thought the voice was familiar, and looked up. To see – Vikki.

'Hold it. You're me!'

'No, *I'm* me. You're you. But you're close. We're both us.'

There was a second Space Hopper too.

'Where have you two come from?'

'No time to explain – you'll find out soon enough. We've brought a portable white hole with us. You can use it to get out.'

With one portable white hole, our heroine was free . . . 'That's *orthog*! You're coming too, of course.'

'No, we can't. It won't work if we do that.'

'What won't work?'

'Causality won't work. Don't try to think about it *now* – there's plenty of time for that later. Just jump into the event horizon of the white hole. Oh, and take it out with you.'

'We can do that?'

'Yes, it's a special model. It shuts itself down automatically after you've gone through, and only opens up when you need it.'

'Then what?'

'You will be contacted,' said the second Space Hopper mysteriously.

•

'Whew!' said the Space Hopper. 'That was a narrow squeak.'

'It certainly was,' said Vikki. 'And also,' she added after a few moments' thought, 'completely mad.'

'Mad?'

'Hopper, *we rescued ourselves.* But to do that, we had to have escaped, and we couldn't because there was no way out!'

'Mmm. But obviously we did escape . . . or else . . .' He stopped. 'I dunno. You're right, it's mad.'

'So now?'

'We hang around and wait. "You will be contacted", so my alter ego told us. So I imagine we—'

'Hi there!' said the Space Hopper. Two Space Hoppers stared at each other. So did two Vikkis. 'Now listen very carefully, I shall say this only—'

'I think you've got some explaining to do. Anyway, how did you two get out of the black hole?'

'Hmm. You're making a rather *big* assumption there. But you're right, we did. We used the time machine, of course.'

'What time machine?'

'This one. Which, by the way, *you* are going to need, so I'll give it to you now. Oh, and keep the portable white hole, you'll need that shortly too.'

Slowly, an *idea* was forming in Vikki's mind. Now that they were out of the black hole, and were in possession of a time machine and an automatically deploying portable white hole . . .

'Hopper! They were *us*!'

'But they said they weren't.'

'Not exactly. They were a second set of us – from the future. Our *own* future. They were a slightly older version of us!'

'That's right,' said the slightly older version of Vikki, 'and do you see what you've got to do?'

'Most of it,' said Vikki. 'But how do we—'

The slightly older version of Vikki told her.

•

Knowing where the black hole was kept, this time round, and having full use of the VUE, it was easy to sneak into the Hawk King's Wormhole Room unnoticed. They had travelled back in time, using the time machine they'd just given themselves, to a moment a few minutes after their first disastrous descent into the black hole.

Now they were going to do it again – but this time they were forewarned, and therefore forearmed. With a time machine and a white hole.

Once inside, they went through a now-familiar routine.

'We're stuck, then,' the original Vikki was saying disconsolately.

'You certainly are,' said Vikki – who was now the slightly older version of herself.

The original Vikki thought the voice was familiar, and looked up. To see – Vikki.

'Hold it. You're me!'

'No, *I'm* me. You're you. But you're close. We're both us.'

The original Vikki saw there was a second Space Hopper too.

'Where have you two come from?'

'No time to explain – you'll find out soon enough. We've brought a portable white hole with us. You can use it to get out.'

With one portable white hole, our heroine was free . . . 'That's *orthog*! You're coming too, of course.'

'No, we can't. It won't work if we do that.'

'What won't work?'

'Causality won't work. Don't try to think about it *now* – there's plenty of time for that later. Just jump into the event horizon of the white hole. Oh, and take it out with you.'

'We can do that?'

'Yes, it's a special model. It shuts itself down automatically after you've gone through, and only opens up when you need it.'

'Then what?'

'You will be contacted,' said the Space Hopper mysteriously.

•

Vikki (slightly older version) and the Space Hopper (slightly older version) watched their younger versions depart, taking the portable white hole with them.

'So how do *we* get out?' asked the Space Hopper. '*They've* got the white hole, so now *we're* stuck here. Do we wait for a *third* version of us to rescue us?'

'Not at all. You're forgetting, *we've* got the time machine.'

'So?'

'All black holes eventually lose their energy by Hawk King radiation, and evaporate.'

Vikki's brain screeched to a halt. 'No, hold it, wait. I thought a black hole gobbles up everything near it.'

'Oops. That's true for a purely relativistic black hole, but not – as the Hawk King discovered – for a quantum one. When quantum effects are taken into account, black holes turn out to be *hot*.'

'Hot? Why should something that nothing can escape from be hot? How does the heat escape, then?'

'It escapes because in quantum physics a vacuum isn't empty.'

'I thought that's what "vacuum" meant.'

'It did, long ago,' said the Space Hopper, 'but not any more. On the smallest of scales, a quantum vacuum is a seething foam of particles and virtual particles, being created in pairs from nothing, coexisting for a split second, and then annihilating each other – or something else. A quantum vacuum is only empty on average. Now, suppose that a particle pair comes into being just outside the event horizon of a black hole. What happens?'

'They both get gobbled up?'

'No. *One* gets gobbled up, but the law of conservation of momentum implies that the other one has to go in the opposite direction, so it escapes. Hawk King radiation is the result of all those escaping particles. So black holes are hot. Over time, they lose more and more heat by radiation, cool down, and eventually evaporate.'

'How much time does it take?' Vikki asked. It affected their current predicament.

'Very quick for atomic-scale black holes, enormously long for bigger ones,' replied the Space Hopper. 'This one will take for ever to evaporate.' He sighed.

'We've *got* for ever,' said Vikki, in a flash of inspiration. 'We have a time machine.' The Space Hopper perked up again. Some careful crosstime planning was in order . . .

And so they fast-forwarded several zillion years, until the black hole evaporated around them and they were floating freely in metaspace. Then they fast-backwarded several zillion years – though not *quite* as many as they had fast-forwarded . . . and again they experienced a familiar conversation from the other side . . .

'So now?'

'We hang around and wait. "You will be contacted", so my alter ego told us. So I imagine we—'

'Hi there!' said the Space Hopper. Two Space Hoppers stared at each other. So did two Vikkis. 'Now listen very carefully, I shall say this only—'

'I think you've got some explaining to do. Anyway, how did you two get out of the black hole?'

'Hmm. You're making a rather *big* assumption there,' said the slightly older version of Vikki. 'But you're right, we did. We used the time machine, of course.'

'What time machine?'

'This one. Which, by the way, *you* are going to need, so I'll give it to you now. Oh, and keep the white hole, you'll need that shortly too.'

Slowly, an *idea* was forming in the mind of the slightly younger version of Vikki. Now that they were out of the black hole, and were in possession of a time machine and an automatically deploying portable White Hole . . .

'Hopper! They were *us*!'

'But they said they weren't.'

'Not exactly. They were a second set of us – from the future. Our *own* future. They were a slightly older version of us!'

'That's right,' said the slightly older version of Vikki, 'and do you see what you've got to do?'

'Most of it,' said the slightly younger version of Vikki. 'But how do we—'

The slightly older version of Vikki told her.

•

'And that's that,' said the Space Hopper.

'Wha— but they've just gone back in time to go into the black hole, and—'

'No, they're earlier versions of us, and we've already been through all that. Think about it: we're out of the black hole, we've done all the time travelling and handing over of technology that was needed to *get* us out—'

'But *where did that technology come from?*'

'The white hole and the time machine? We gave it to us.'

'But—'

'You asked the wrong question. *Where did that technology go to?* We haven't got it now, have we?'

'No.'

'So that's all right. The Mathiverse hasn't gained anything it didn't have before.'

'But it's mad,' said Vikki. 'The causality doesn't work.'

'On the contrary,' said the Space Hopper. 'The causality is the one thing that *does* work! You're just not thinking about it the right way.'

'I don't know how else to think about it!' Vikki wailed.

'Use a Feynman diagram, of course', said the Space Hopper.

'What's that when it's at home?'

'Richardfeynman was a Planiturthian who was doodling pictures of world-lines of particles moving through space and time, when he suddenly realized that antiparticles moving forwards in time could be viewed as ordinary particles moving *backwards* in time.

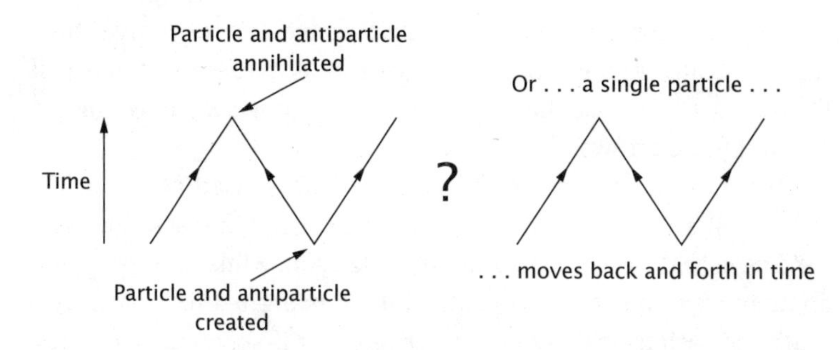

And then he noticed that there was just one line in his picture, whizzing back and forth through time, and suddenly a really revolutionary idea popped into his head. 'Why are all electrons identical?'

'Dunno.'

'Because they're all the *same* electron! It just keeps shuttling up and down in time, and each new incarnation turns it into what *seems* like a different electron.'

'Cunning.'

'Very. And possibly wrong – nobody knows for sure. But the point is that by drawing a picture of the world-lines, you can sort out the subjective sequence of events from the objective one, and check that the causality works.' And the Space Hopper quickly used the VUE to sketch a Feynman diagram of their adventures down the Black Hole.

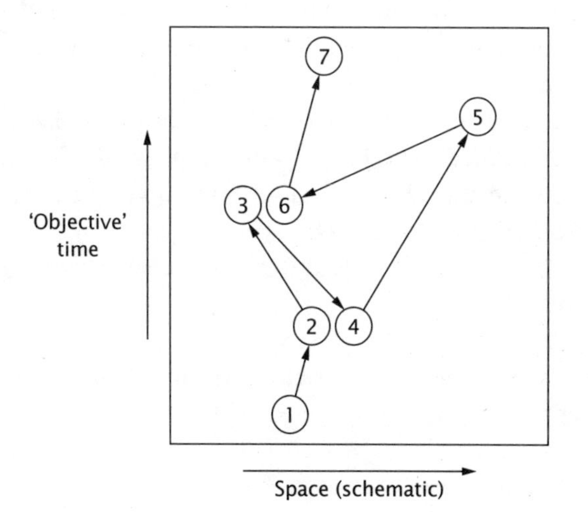

'OK,' said the Space Hopper. 'The Mathiverse's "objective" time runs vertically up the picture. The horizontal space direction is just there to separate out the main events. And our *subjective* time is shown by the arrows.

'Let me run through what happened. We started at (1), in the Hawk King's Palace. Then we got trapped inside a black hole that we'd been told was a wormhole – that's (2). And while we were there, another we came back from the future using a time machine – that's the path from (3) to (4) – and dropped ourselves back into the black hole to rescue ourselves. We also brought a white hole to create an exit. Where did we get it, and the time machine? That's another story – I'll come back to it once we've sorted out the main timeline.

'Anyway, from (2) we used the white hole to get ourselves to (3), back into ordinary space. We left the time machine with the "other" Vikki and Space Hopper inside the black hole and proceeded into the future in the normal way. Then we went from (3) to (4) – I've already told you that bit. After that we used the time machine to go zillions of years into the future, to (5), where the black hole evaporated and we were free of its event horizon. We promptly used the time machine *again* to go slightly fewer zillions of years into the past, to (6). At that point we met our former selves (3) coming the other way – in time, that is – and gave them the hardware they needed to go back to (4). Then we continued from (6) towards the distant future (7), using the normal method of waiting for time to pass of its own accord.'

Vikki ran her eyes over the Feynman diagram, checking each event. It all seemed to fit. But . . . 'I'm still puzzled,' she said.

'Do you see that the passage of subjective time, for us, just moved from (1) to (2) to (3) to (4) to (5) to (6), and now we're heading towards (7)? It's just a single linear chain of experience – for us. Subjectively.'

'*Yeeesss . . .*'

'And each event has the right versions of us, and the right equipment?'

'*Yeeesss . . .*'

'Then I don't see what your problem is.'

'Space Hopper, where did the time machine come from? Where did the white hole come from?'

'Where have they gone? That's what you ought to be asking.'

'I don't know! We're *out*, and we don't have them any more . . . but I can't quite see how the trick worked.'

'Actually, the question is not *where* did they come from, but *when*. Trace the temporal movement of the white hole.'

'Um. We picked it up at (2), carried it to (3), took it back in time to (4) – oh! And then we handed it over to our former selves at (2) again!'

'That's right. The portable white hole is going – was going? Will go? Will have being gone? – round and round a closed timelike curve. A time loop. Its past runs into its future: it has always been in the time loop and it always will be. And what about the time machine?'

'We were first given that – in our own subjective sequence of events – at (3). We took it back to (4), used it to get to (5) without tottering off this mortal coil, used it again to get us back to (6) – and then handed it back to ourselves at (4). The *time machine* is going round a time loop too!'

'Yes. But a different one from the loop that the white hole is in.'

'Right. And because they're in time loops, but we're in an ordinary open-ended chain, we emerge from the whole process without any fancy technology that we couldn't possibly possess.'

'Precisely. All of which goes to show—'

'—that causality is totally *weird* when you've got a time machine. Even if it disappears up its own – er – closed timelike curve.'

'It's only weird *to us*. As far as the Mathiverse is concerned, it all makes perfectly good sense. It seems weird *globally* – viewed as a whole. But each piece of it is entirely reasonable and logical. It's *easy* to explain where the time machine and the white hole come from. They've always been stuck in their time loops, and always will be. The hard thing to explain, Vikki, is where *we* come from. Because that seems to need an infinitely long chain of causality. Like you said, *our* timelines are open-ended.'

That reminded Vikki of something. 'Yes, maybe . . . but Space

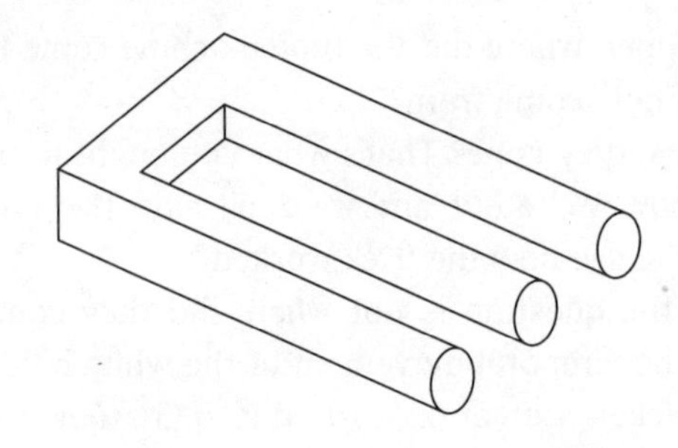

Hopper,' she said, 'do you recall that funny picture that Spacelanders sometimes use to baffle themselves with?' She fiddled with her VUE, and a line drawing of a solid object appeared:

'Yes, I remember. The object looks OK, but it can't actually exist.'

'My point exactly. Now, what do you mean by "look OK"? You mean that each piece is entirely reasonable and logical. And when you say "can't actually exist", you're saying that—'

'It seems weird globally.'

'So maybe the kind of twisted causality that we think we've just experienced can't *really* happen in a relativistic spacetime,' said Vikki. 'We're missing a global law of nature that could rule it out.'

'Such as?'

'Such as having to avoid the Great-great-grandfather Paradox.'

The Space Hopper gave this due consideration. 'You could be right,' he said, 'though there are other ways to get round the Great-great-grandfather Paradox. But if you *are* right, how come we got out of the black hole?'

Vikki had no answer to that.

•

Thanks to the logic, or illogic, of causality in closed timelike curves, Vikki and the Space Hopper were once more free to roam the Mathiverse. But their plan to use a time machine to help the Paradox Twins had to be abandoned, for they dared not return to the Domain of the Hawk King. The twins would just have to pursue a career of hiring themselves out to writers of physics texts.

Despite their narrow brush with death, however, Vikki still wanted to understand more about wormholes and time travel. The whole topic was just too interesting for her experiences to put her off thinking about it.

'The Hawk King's heavy engineers must have some amazing construction methods to be able to *build* wormholes to order,' she said. 'I'd like to know just what's involved in that.'

The Space Hopper jiggled his antennae. 'Well, we can't return to the Hawk King's territory, that's for sure. So we won't be able to *watch* a wormhole being made. But there's another part of the Mathiverse where people think about that kind of problem but don't actually *make* anything to put their thoughts into action. I suppose we could go there if you want.'

'If it's not too much trouble.'

'No more than usual. Anyway, I haven't met up with the Charming Construction Entity for a while. Be nice to find out what he's up to these days.'

'That's a very curious name.'

'It was a mistranslation. He was originally known as Civil Engineer. When he emigrated to his current part of the Mathiverse, he didn't speak the language very well, so he used a dictionary to translate. Unfortunately he chose the wrong meaning for "civil", and a rather roundabout phrase for "engineer".'

'Why didn't he go back to the right translation when he found out?'

'Marketing. His customers were mostly romantics who were uneasy about technology. They found the mistranslated name less threatening.'

The Charming Construction Entity can best be described as a land octopus. It had four stubby legs and eight long, flexible tentacles. If it had ever got round to building any of its designs, the tentacles would have come in useful, but even at the theory and design level they gave it a major advantage over its competitors. When Vikki and the Space Hopper arrived in its metaspatial office, it was building a model of a spinning black hole with two of its tentacles, answering the phone with a third, drawing two separate blueprints with two more, photocopying a letter with the sixth, signing contracts with a seventh, and stirring its tea with the tip of the eighth.

When the Charming Construction Entity saw it had visitors, it dropped all these activities and slumped into a comfortable armchair. With eight arm-rests, forming an octagon. Vikki looked wistfully at the phone, but she knew that there was no metaspatial connection to Flatland's rudimentary telephone system.

'I'm not surprised you had trouble with the Hawk King's wormhole,' said the Charming Construction Entity. 'They're unstable things at the best of times. I spend most of my time trying to find improvements.'

'How does he build his wormholes, anyway?' Vikki asked.

'It's not easy. There are several serious technical obstacles. The worst one is the question of whether users can really get through the wormhole. It's not so hard to build the wormhole and move its ends around. That's just a matter of creating intense gravitational fields, which is the Hawk King's forte. The main problem is what you might call the "catflap effect". With the earliest designs, when you traversed the wormhole it would shut on your tail.'

'Why not go through faster?'

'It turned out that, to get through without getting your tail trapped, you'd need to travel faster than light, so that was no good.'

'How does that come about?'

The Charming Construction Entity grubbed around among its papers and tossed one across to her. 'Take a look at the spacetime geometry in this Penrose map.'

'Named after a Planiturthinan named Rogerpenrose,' the Space Hopper added helpfully. 'Not because it's in rows and drawn with a pen.'

Vikki ignored him and stared at the cabalistic scrawls. 'I don't know how to read this.'

'Then I'll tell you', said the Charming Construction Entity. 'You know that when you draw a map of a curved space – a sphere, say – on a flat sheet of paper, you have to distort the coordinates.'

'Sure. Planiturthians suffer from that problem because they live on a sphere. They have to compromise, by drawing curved lines of longitude, say. They can choose to preserve some features – lines of latitude, or directions, or areas. But not all of them.'

'That,' said the Space Hopper, 'happens because curvature is an invariant – you cannot change it. A plane has zero curvature, whereas a sphere has positive curvature. That *proves* you can't preserve all features of a spherical surface in a planar map.'

'Thank you for that,' said the Charming Construction Entity. 'It's the same for spacetimes. The Penrose map of a spacetime also distorts the coordinates – but in a way that doesn't change light cones. They still run at 45 degree angles. Now, Vikki, this is a Penrose map of a wormhole. Any timelike path that starts at the wormhole entrance, such as the wiggly line shown, must run into the future singularity. There's no way to get across to the exit without going outside the 45 degree angle limit – without exceeding the speed of light.'

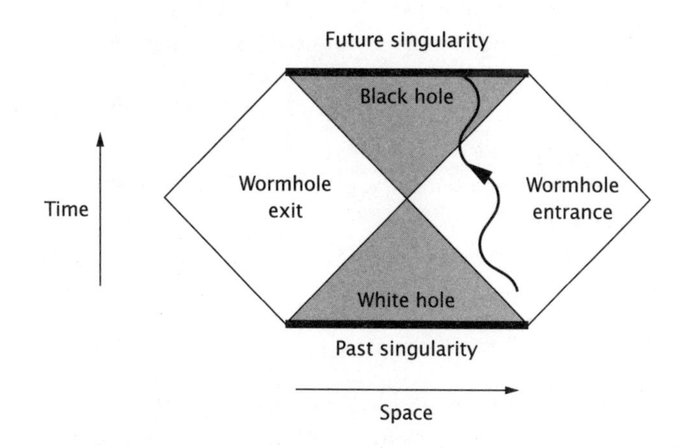

•

Threeday 25 Noctember 2099

This wormhole stuff is a gripping tale. But I'm feeling rather upset right now, Diary Dearest.

You see, I've just realized: back in Flatland, today is Crisp Moose Day. Traditionally, a time when all families get together to celebrate over a special seasonal meal. And this year it's extra-special, because 2099 is the last year of the century!

Now here's a minor curiosity. You know that the Flatland

calendar and the Planiturthian one coincidentally move in step, except that ours adds 100 to theirs? Well, this isn't *quite* true. The Space Hopper tells me that in the Planiturthian calendar, new centuries *end* with years numbered 00 and start with years numbered 01. Whereas on Flatland new centuries *end* with years numbered 99 and start with years numbered 00. However, it turns out that most Planiturthians don't know how their own calendar works, so most of them celebrate new centuries one year too early. You can understand why: our system is much less confusing and far more logical.

Sorry, Diary. I've been rabbiting on about the numerology of new centuries in an effort to forget how miserable I'm feeling today. It was all right until I became aware of the date. The rest of my family will be sitting round the big table waiting for Mum to finish crisping the Moose . . . and it will all be spoilt because I'm not there, and they'll be worrying about what's happening to me . . . and I haven't even sent them a crisp moose card!

Now *there's* an idea . . .

Flatland telephones may be incompatible with metaspace, but Flatland *cards*?

I'll ask the Space Hopper whether something can be done.

•

The smell of crisping moose filled the old pentagonal house. Les and Berkeley were charging round like mad things and playing with their new toys; cards from friends, relatives, and acquaintances had been collected in one corner; and the natural logs in the hearth were burning brightly. But the toys just reminded Grosvenor of Vikki's childhood, the cards reminded him that she hadn't sent one, and the fire reminded him of how he had burned Albert's book – an act that he still was convinced had played some part in Vikki's dramatic disappearance.

No doubt Lee was feeling much the same. It was putting a real damper on the proceedings, and the boys would soon notice. He needed something to cheer himself up. So he decided to pour himself and Lee a couple of drinks.

He always kept the drinks cabinet locked, ever since Lester had got hold of a bottle of torquila and been found asleep in the washing basket wearing next door's dustbin lid. He got out the key and opened the cabinet . . .

There was an envelope inside. Addressed to him and Jubilee.

His vertices tingling, he picked it up and took it through to the kitchen, unopened. His wife was busy checking whether the moose was crisp enough yet.

'Lee, have you been putting things in the drinks cabinet?'

'No, dear – what kind of things, dear?'

'An envelope. This one.'

'Nothing to do with me, dear.'

'Well, it's got your name on it.'

Jubilee put down the moose-strainer and came over to take a look. 'It's got *your* name on it too, dear. And it's – Grosvenor, that's *Vikki's* writing!'

'Yes, I wondered about that. It's certainly very similar . . .'

'Why don't you just open it and find out?'

'I'm . . . Lee, I'm scared of being disappointed.'

Jubilee took the envelope from him and ripped it open. 'Don't be silly. Look, it's a card.' She opened the card, which read: 'Wishing you a Very Crisp Moose'. There was also a note – a very long note, by the look of it.

She unreeled the paper, and they began to read.

•

Vikki felt a lot better once the Space Hopper had agreed that she could use the VUE to deposit a Crisp Moose card in the plane of Flatland. It had been his idea to put it somewhere that only Lee or Grosvenor would have access to: that fact alone would support the story that she had insisted on including. She had thought of the drinks cabinet – on Crisp Moose Day, her dad was bound to open it.

To take her mind off what might happen when her father read it, Vikki was badgering the Charming Construction Entity for more details about time machines. 'How does the Hawk King's wormhole work, then?'

'Maybe it doesn't. Maybe it's all marketing propaganda. You didn't actually pass *through* a wormhole, did you?'

'True. But surely customers would catch on if the things never worked.'

'Possibly. But you met the Hawk King. Would they dare say so in public? However, I think that he does have working wormholes. His methods are commercial secrets, of course, so I can't be sure, but I've done a lot of calculations and I think I've figured out how the trick works. I reckon he must be threading his wormholes with exotic matter.'

'That sounds . . . well, exotic. What is it?'

'It's an unorthodox form of matter that exerts enormous negative pressure, like a stretched spring.'

'Antimatter?'

'No, that's different,' the Space Hopper interjected. 'When matter meets antimatter they annihilate in a blaze of energy. Exotic matter just pushes instead of pulling.'

'Antigravity, then.'

'Well . . . technically, no. Sort of, though.'

'Anyway, exotic matter is the most obvious way to hold the catflap open so it doesn't trap your tail. But there's a more old-fashioned method that cuts out the need for exotic matter altogether. And because it doesn't involve *building* a wormhole, there's no time-barrier effect. You can go back to any time you want. Depending on what nature has up her sleeve.'

'I don't follow you,' said Vikki.

'I'm talking about using a naturally occurring time machine. A *rotating* black hole, formed when a rotating star collapses gravitationally. The Schwarzschild solution of Einstein's equations – you know about that?'

'Yes, the Space Hopper told me.'

'Well, that corresponds to a *static* black hole, formed by the collapse of a nonrotating star. In 1962, a Planiturthian called Roykerr solved the Einstein equations for a rotating Black Hole, now known as a *Kerr black hole.*'

'The Planiturthians know about two other kinds of black hole,' the Space Hopper said. 'There's the Reissnernordström black hole, which is static but has electric charge, and the Kerrnewman black hole, which rotates and has electric charge. It is almost a miracle that an explicit formula for the solution exists – and it was definitely a miracle that Roykerr was able to find it. It's extremely complicated and not *at all* obvious.'

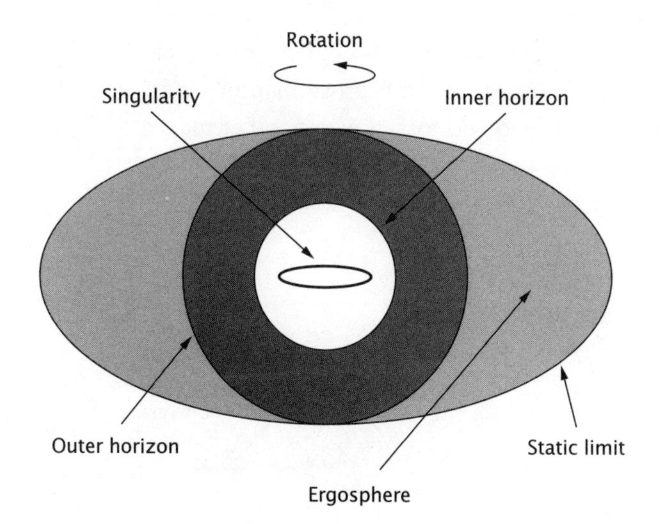

'But it has spectacular consequences,' said the Space Hopper. 'One is that there is no longer a point singularity inside the black hole. Instead, there is a circular ring singularity, in the plane of rotation. The Charming Construction Entity's Penrose map shows that any matter entering a static black hole has to fall into the singularity – but in a rotating one, it need not. It can either travel above the equatorial plane, or pass through the ring.'

'Yes, and a rotating black hole's event horizon also becomes more complex. In fact, it splits into two. Signals or matter that penetrate the *outer horizon* can't get back out again; signals or matter emitted by the singularity itself can't travel past the *inner horizon.* Further out still, but tangent to the outer horizon at the poles, is the *static limit.* Outside this, particles can move at will. Inside it, they have to rotate in the same direction as the black hole, although they can still escape by moving radially.'

'Yes, yes!' said the Space Hopper, getting all excited. He loved this kind of complication. 'And between the static limit and the outer horizon is the *ergosphere.* If you fire a projectile into the ergosphere, and split it into two pieces, one being captured by the black hole and one escaping, then you can extract some of the black hole's rotational energy.'

'The most spectacular consequence of all,' said the Charming Construction Entity, 'is the Penrose map of a Kerr black hole.' He rummaged around and tossed over another map.

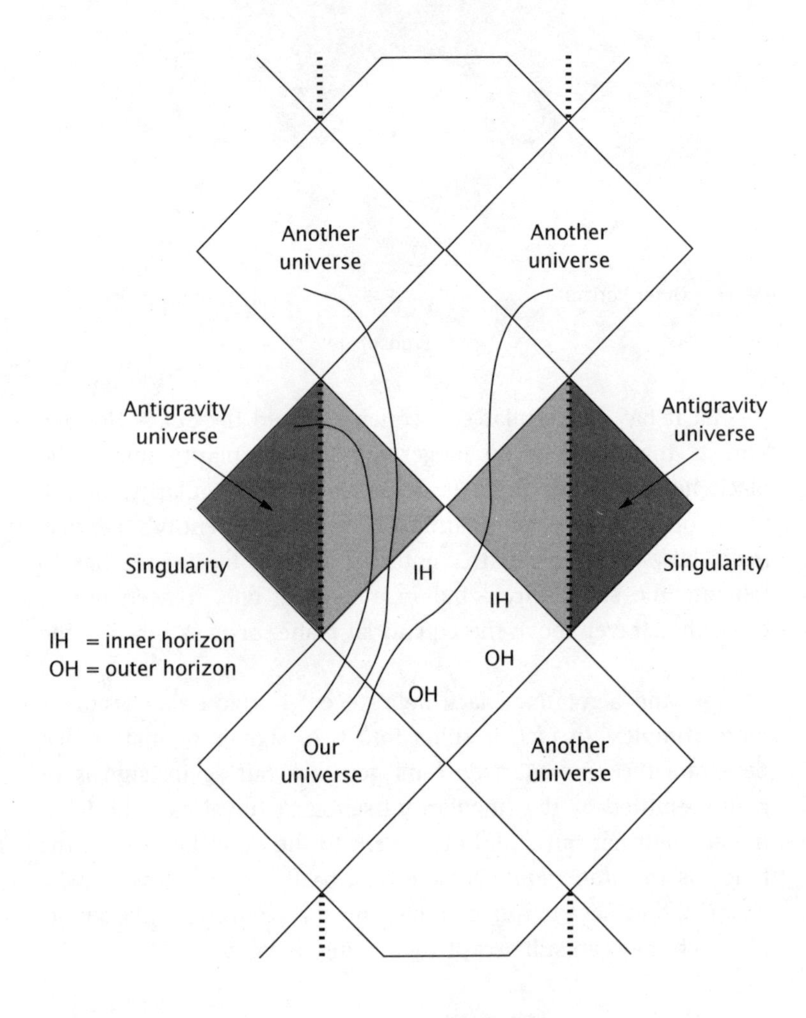

'The white diamonds represent asymptotically flat regions of spacetime – one in the Planiturthian universe, say, and several others that might also be in that universe, but could be in totally different ones. I've drawn the singularity with broken lines, because it's possible to pass through it.'

'How?'

'By diving right through the ring, Vikki.'

'And now comes the fun bit,' the Space Hopper butted in. Beyond the—'

'No, I want to tell this bit. Beyond the singularities lie antigravity universes in which distances are negative and matter repels other matter. In those regions, any body made from normal matter will be flung away from the singularity to infinite distances. Now, Vikki, I've drawn several legal trajectories – ones that stay below lightspeed – as curved paths. As you can see, they lead through the wormhole to any of its alternative exits.'

'The most spectacular feature of all,' said the Space Hopper, attempting to reclaim his part in the conversation, 'is that this is only part of what's going on. The complete diagram repeats the same pattern indefinitely, so there are an infinite number of possible entrances and exits!'

Vikki had to admit she was impressed. But what was all this *for*?

'Well, if you used a rotating black hole instead of a wormhole, and towed its entrances and exits around at nearly lightspeed with the Hawk King's matter-processing equipment, you'd get a much more practical time machine – one that you could get through without running into the singularity or getting your tail trapped.'

'It all looks very complicated.'

'Oh,' said the Charming Construction Entity, deflated by her lack of enthusiasm. 'Well, if you don't fancy trying to control Kerr black holes, you could settle for a much simpler kind of singularity: *cosmic string*. That gives a static spacetime.' And he dug out yet another map.

'Let me run through the details for you. The best way to visualize cosmic string is to use two dimensions of space—'

'Like Minny Space?'

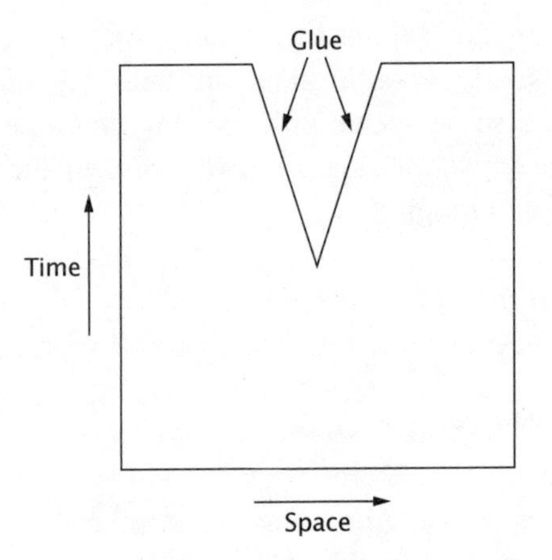

'Oh, you've met the Space Girls? Congratulations on your survival skills. Yes, just like Minny Space, but with some parts snipped out. You have to cut out a wedge-shaped sector and glue the edges together.'

'Like Squarey Space? You don't mean real glue, do you? Squarey was a square whose edges were "glued" together. Conceptually.'

'That's exactly right. But in this case you can actually do the gluing if you insist on it. And it may help you visualize the result. If you make a model out of paper, and really do cut out wedges and glue their edges together, you end up with a pointed cone. But you're right: mathematically you can just identify the corresponding edges without doing any bending or using any actual glue.'

'Wow! Coney Space.'

'Funny you should say that,' the Charming Construction Entity said. 'I hear that one of the Space Girls is leaving and there could be a place for a new member . . . but let's not get distracted. The time coordinate works just as it does in Minkowski spacetime – and to get the right shape for light cones you really should just identify the edges and not make actual cones, OK? Now, throw in a third space coordinate and repeat the same construction on every perpendicular cross-section, and you'll get a fully fledged cosmic string. It behaves like a line mass – nonzero mass concentrated on a line. To make a model of one of those, you can thread lots of identical cones on a length of – well, string. Each cone is a spacelike section of the actual spacetime.'

'So what does that do for you?'

'Something very subtle and interesting. A cosmic string has a mass, which is proportional to the angle of the sector that you cut out. But it doesn't *behave* like an ordinary mass. Everywhere except at the tips of the cones, spacetime is locally flat – just like Minkowski spacetime. The apparent curvature of a real cone is "harmless" – yes, Vikki, just as Squarey Space wasn't really curved even though she was a torus. In a cosmic string there is some genuine curvature – which in relativistic gravitation is equated to mass – but what it does is create *global* changes in the spacetime topology. These global changes affect the large-scale structure of geodesics – particle paths. But not the small-scale local structure. If you were standing next to a massive cosmic string, you wouldn't feel any gravitational attraction.'

'Then how would you know it was there?'

'By looking for global effects – things happening over greater distances', said the Charming Construction Entity. 'For instance, matter or light that goes past a cosmic string is gravitationally lensed.'

'Oh, I know about that. The mass bends light. But how can it do that, if the light doesn't "feel" the gravity?'

'The bending is caused by the global form of the spacetime, not by a loss of local flatness. Geodesics on a paper cone correspond to straight lines on the flat paper, but when you bend the paper to make a cone, those lines get "bent" too – in the sense that what were originally parallel lines can converge. Let me draw you some geodesics, and you'll see what I mean.'

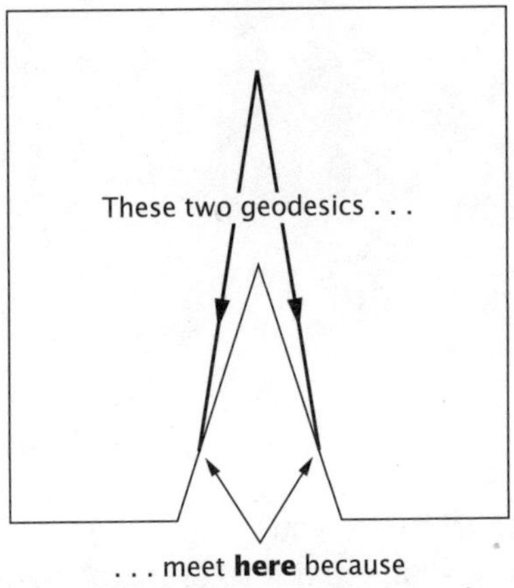

'For time travel purposes,' the Charming Construction Entity continued, 'a cosmic string is much like a wormhole because the mathematical glue lets you "jump across" the sector of Minkowski spacetime that you've cut out. Way back in 1991 a Planiturthian named Richardgott exploited this analogy to design a time machine. He showed that the spacetime formed by two cosmic strings that are whizzing past each other at nearly lightspeed contains closed timelike curves.'

'Fair enough, but could you actually make a time machine that worked that way?' asked Vikki.

'A shrewd question.'

'Yes, she's learning,' the Space Hopper said with pride.

'Well . . . in 1992 by the Planiturthian calendar, Seancarroll, Edwardfarhi, and Alanguth proved that there isn't enough available energy in the Planiturthian universe to *build* a Gott time machine', the Charming Construction Entity answered. 'More precisely, that universe never contains enough matter to provide such energy from the decay products of stationary particles.'

'Oh. Pity!'

'Though if the Planiturthians could develop a sufficiently powerful new energy source . . . but I'm afraid that's not in the works yet. However, I recall that surveys of the distribution of galaxies in their universe has revealed that they clump on vast scales, forming structures hundreds of millions of light-years long. This clumpiness is too great to have been caused by gravitational attraction among the known matter.'

'So?'

'One theory is that the clumps were seeded by naturally occurring cosmic strings,' said the Charming Construction Entity. 'Not by gravitational attraction, of course, because there isn't any, but by unexpected convergence of geodesics. If so, those cosmic strings ought still to be around. So maybe, just maybe . . .'

'The Planiturthians really could build themselves a cosmic string time machine,' the Space Hopper finished for him, 'which would be nice.'

'But what about the Great-great-grandfather Paradox?' asked Vikki.

'Ah, that. Yes.' The Charming Construction Entity always found such objections tiresome. 'If you can *make* a time machine, then presumably the paradoxes will iron themselves out.'

Vikki didn't find this entirely convincing. 'Maybe they won't, though. Maybe when you build a time machine the logic of your universe unravels and it comes to pieces. Or maybe history becomes something that can be changed, and you get huge Time Wars as people range up and down history trying to alter it to suit their own wishes.'

'Could be,' said the Space Hopper. 'On the other hand, every time you change history, you might get switched to a different timeline in an alternative universe. So you'd think history had been changed, when actually the original timeline was still present, elsewhere in the Mathiverse. No paradoxes, then.'

'That sounds suspiciously like a cop-out to me,' said Vikki.

'Not at all,' said the Space Hopper. 'It's one way to make sense of quantum indeterminacy.' And before she could protest, the VUE once more took control of her senses.

•

'It's like this,' said the Space Hopper. 'Despite what everyone seems to think, quantum superpositions don't tell you anything very interesting about cats. But they could be important for time travel, by resolving the Great-great-grandfather Paradox.'

Vikki wished he hadn't reminded her of Albert. Visions of Crisp Moose Day filled her head, and for a moment she felt dizzy. To distract herself, she asked, 'How could they do that?'

The VUEfield shimmered . . . Flatland was below them. If anything, that made her feel even worse, but she struggled against a rising tide of emotion. The Space Hopper, who wasn't exactly tactful at the best of times, didn't notice.

'Is that—?'

'Your ancestor Albert? Only in VUE simulation. So don't be upset by anything that happens, OK? Promise?'

'Promise,' Vikki snivelled. 'Oh, look, he's in prison, poor thing – you can see the dots over the window. And he's writing something. Is it *Flatland*?'

'The very same. Now, let's run time back to before he was imprisoned. When the Sphere was about to visit him. That's what started it all.'

'Oh, there he is in his house. And – I can see the Sphere rising up towards the plane of Flatland!'

'Yes. What would you do to stop Ancestor Albert encountering the Sphere?'

'I'd . . . I'd turn up just before and find some excuse to get him out of the way.'

'And that's just what you *are* doing. Look, he's got up and gone to the door.'

'Yes, but – he's stayed behind as well! And both versions of him are semitransparent!'

'Yes. His universe has split into a quantum superposition of two different versions. In one, you managed to distract him; in the other, you didn't. From now on, both universes will follow distinct paths.'

They could hear the conversation in Universe-1, faintly. The time-travelling Vikki was pretending to be collecting money on behalf of a charity. The Sphere materialized in Albert's room – but the room was empty. Puzzled, the Sphere pottered around for a bit, and then disappeared again.

In Universe-2, the Sphere materialized in Albert's room, confronted Albert – and at that moment his eventual imprisonment became inevitable.

'OK,' said the Space Hopper, 'Universe-2 is the one in which Albert went to prison, and therefore it's the universe that you were in when you used the time machine to go back to distract him. Universe-1 is what happens instead, once you *have* distracted him. Let's fast-forward to the moment when you climbed into the time machine. Got it. Here you are, in Universe-2 . . . you get into the time machine, go back, still in Universe-2 . . . until eventually you're back before the time when Universes 1 and 2 split. Now, in Universe-1 you *do* manage to distract him . . . which means that in *that* version of the universe there was no need for you to go back in a time machine to stop him encountering the Sphere. Because he never did anyway, right?'

'Right.'

'So do we see you get into a time machine in Universe-1?'

'No. Only in Universe-2.'

'Exactly! And when you come back from time travelling, you remain in Universe-2 because that's where you came from. And Universe-2 is the one where he did go to prison and you did need to go back and try to distract him. So it's all consistent. More generally, if the universe splits into alternatives every time some decision could or could not be made, then time travel would just switch you between alternatives, maybe creating new ones in the process or changing their probabilities. But there wouldn't be any time travel paradoxes.'

Vikki thought about this, and spotted what seemed to be a major flaw. 'I thought you kept saying that real cats don't superimpose. How come real *universes* do?'

'Ah. They don't: they become alternatives. The people in them perceive one or the other, never both. But in terms of quantum theory, the physics makes more sense if you pretend that both possibilities coexist. It's a way to represent quantum uncertainty and make it mathematically respectable. Planiturthians call it the "many worlds" interpretation of quantum theory.'

'Is the Planiturthian universe *really* a branching network of alternative possibilities, then?' asked Vikki.

'That's certainly a consistent interpretation of the mathematics,' said the Space Hopper. 'The question is, does every consistent interpretation of physicists' mathematics represent reality?' There was a long silence.

'Well, does it?'

'You'll have to work out your own VUEs about that.'

WHAT SHAPE IS THE UNIVERSE?

Another day, another trip through the Mathiverse. The sphere of Planiturth hovered before them – dazzling ultramarine, smeared with white, patched with greens and browns.

'How anyone could ever have doubted that the Planiturthian universe is based on geometry is beyond me,' said the Space Hopper. 'Just look at the thing they're *living* on!'

'I guess they were too close to it to understand what it was,' suggested Vikki.

'You're learning, aren't you?' said the Space Hopper, impressed.

'Not to go by appearances?'

'Exactly. You've put your tip right on it. Yes, it often helps to take a step back and see the whole more clearly, as well as to take a really close-up look at all the nuts and bolts . . . Speaking of which, we've had a very close-up look indeed at the small-scale structure of the quantum world, but we haven't yet—'

The jewelled topaz sphere receded as if snatched away. Its diminutive, pockmarked satellite sped past them, vanishing to a speck. The entire Planiturthian solar system turned into a child's toy and disappeared against the velvet backdrop of stars.

The stars radiated away as if the universe were contracting to a point.

The speckled dust of pinpoint lights merged, acquired form – a whirlpool of fuzzy light. The sedately twirling spiral arms of the Home Galaxy.

This was just the beginning. Now the whole process repeated on a galactic level. The heavens were filled with luminous skeins, cosmic cobwebs millions of light-years across, now shrunk to the size of a paper doily.

'What's happening?' asked Vikki, breathless at the sheer *pace* of the motion.

'We're taking a step back. Looking at the cosmos *as a whole*.'

'And what do we expect to see?'

'The shape of the universe. And its origins.'

Now the contraction had ceased. The Planiturthian universe hung before them, a dark smudge visible only because metaspace has no existence at all and therefore cannot be any particular colour, not even black. The smudge was fuzzy, inchoate, and impossible to see all at once.

The Space Hopper sighed, and a message wrote itself across the formless smudge of the Planiturthian universe:

OUT-OF-INFORMATION-ERROR

'As I thought,' said the Space Hopper. 'The VUE can only show what it's programmed to. It shows Virtual Unrealities, not Realities. It doesn't know what shape the Planiturthian universe should be, so it can't show us any detail. But there's one thing it *does* know. Look at the smudge – what's it doing?'

'I think . . . it's getting bigger! Yes, it is! Or are we getting closer to it?'

'No, it's getting closer to us. It's *expanding*. I don't mean that the stars are moving away from one another through space – I mean that *space* is moving away from itself. Growing of its own accord.'

'Obviously somebody decided that the Planiturthian universe needed more space,' said Vikki, 'but how do you know it's expanding if you don't know what shape it is?'

The Space Hopper's grin was a big ∪ as the smudge swelled and enveloped them.

•

They orbited a pale blue galaxy.

'Oh! It's *beautiful*!' cried Vikki, entranced.

'Well, I chose this one because it's particularly pretty,' said the Space Hopper. 'And it's out near the edge of the universe. But the main thing I want you to remember is its colour.'

'Blue,' said Vikki. 'A wonderful, delicate, heartbreaking blue.'

'It's a very special blue,' said the Space Hopper. 'This is the only monochromatic galaxy in the universe. Meaning that the light that it emits is *all of one wavelength.* It's been like that since it first appeared, and it's always been the only one. Remember that. Because now we're going to travel to the centre of the universe and look back at this galaxy.'

Space surged past them, too fast for the eye to follow; then it stopped.

'Turn around, look back the way we came, and use the VUE's zoom facility to locate that pale blue galaxy,' the Space Hopper instructed her. 'Set the VUE for monochromatic light; then you won't be distracted by billions of other galaxies.'

Vikki searched, using higher and higher magnification, but she couldn't see the blue galaxy. In fact, in all the universe the VUE could detect only one monochromatic galaxy. But that one was red.

'I've found a red one,' she told the Space Hopper.

'Not blue?'

'No. But didn't you say the blue one was the only monochromatic galaxy in the universe?'

'I did. And I was right.'

'Then where did this red one come from? And where's the blue one gone?'

'You tell me!'

•

Fiveday 27 Noctember 2099

The Space Hopper is *such* a tiresome creature! Try as I might, I couldn't puzzle it out. 'Follow the logic,' he told me. But I couldn't see any logic to follow. All I could see, Diary my Sweet, was a red galaxy when I was expecting a blue one.

'You mean this red thing *is* the blue one?' I eventually asked him. 'It's changed colour?'

'That's a very interesting question,' he replied, in that irritatingly smug way he has.

Eventually I figured it out. Since the blue galaxy was the only monochromatic one in the universe – *and* the only one there had ever been, which turns out to be important – and I was LOOKING at a monochromatic galaxy, it had to be the blue one.

The fact that it was red didn't change that conclusion, you appreciate.

So . . . it must have *changed colour*, right?

Sort of.

My first thought was that, seeing as the blue galaxy was at the edge of the universe, and we were in the middle, then it had to be an awfully long way away. And since light can cover only 300,000 kilometres in a second, the light that was reaching us must have started out billions of years ago.

So, of course, the blue galaxy had actually been red, *then.*

Not at all.

'Blue from the day it was born,' the Space Hopper insisted. So I sat down and thought very hard. According to the Space Hopper, that galaxy was blue when the light left it. So the light must have changed along the way!

Nope.

Nothing had happened to the light. Not *as such.*

Then I remembered the Light Brigade. You'll recall that when they first overtook our Universal Touring Machine, when we were cruising along Continuum Carriageway, the siren made a noise like this:

WHEEEEEEEEEEEEEEEEEEEEOOOOOoooooooooooooooooooo

Now, it may not have been obvious at the time, but the EEEE part was a *higher pitch* than the OOOO part. This, the Space Hopper had told me, was caused by the Doppler effect. (Really, Diary, the next bit ought to be done relativistically, but the calculations are messy and the answer is qualitatively – though not quantitatively – the same.) Because the Light Engine was moving, it was catching up with the sound waves its siren was emitting in the forward direction, so anyone listening to them received *more* waves per second than they would have done if the Engine had been stationary. Higher frequency, right? And that pushed the pitch UP. When the Light Engine had overtaken us, it was getting away from the sound waves its siren was emitting in the backward direction, so anyone listening to them received *fewer* waves per second than they would have done if the Engine had been stationary. Lower frequency, you see. And that pushed the pitch DOWN. So the sound started out as an EEEE and ended up as an OOOO, when really it had been an in-between IIII all along.

But this was the *Light* Brigade, right? It had a flashing blue light on top too, didn't it? As they do. BUT – and I regret I hadn't spotted this at the time, or I would have told you – as it sped away from us, that light looked RED. It only looked blue when the Light Engine was catching up with us.

Now, all you need to complete the puzzle, Diary, is the information that red light has a lower frequency than blue light.

Yes, that's right. Diary, you are a veritable GENIUS. The Light Engine's light had undergone the Doppler effect too. So whenever you see something that ought to be blue, but looks red, it must be moving *away* from you.

Let me rephrase that. My adventures with the Space Hopper have honed my critical faculties – which is to say that I'm not impressed by oxagondung any more. It's not really the *colour* of a moving object that changes. What happens is that all the *frequencies* of the light that emanates from it have to shift a bit. Now, even if the blue light shifts down into the red part of the spectrum, the object may still *look* blue, because the ultraviolet has also shifted down into the blue part of the spectrum. If it's *emitting* any ultraviolet, of course.

And that's why it was so important that the blue galaxy was monochromatic. If *it* looked red, then the frequency really must have shifted – so it had to be moving. Away from the centre.

You can look at lots of galaxies, and although none of the others are monochromatic, you can use 'spectral lines' to tell whether the frequency has shifted. And what you find is this: they're *all* moving away from you! Which means they are also all moving away from one another. It's as if they're dots on a balloon and someone is blowing the balloon up. Except that this is a balloon with a 3D skin.

And that means—

•

'—That the universe is expanding,' the Space Hopper agreed. 'And since we're pretty sure that all of the space that exists contains much the same proportion of galaxies, the only way that can happen is if space itself is expanding.'

Vikki gave this information careful consideration. 'Well, Hopper, I'm not sure that's really a surprise. Most creatures get bigger as they grow up. So presumably it's telling us that the universe is some kind of gigantic organism.'

'Wha—?' The Space Hopper was outraged.

'Only thing is, *what does it eat*? Maybe it eats *time*. It gobbles up the past, and that's what causes everything to move into the future!'

'Vikki, that's the silliest theory I've heard in a . . . well, maybe not . . . it might just *possibly* be—'

'Hopper, it was a joke.'

'Many a good theory starts as a joke. Most end that way, too. But I think I ought to warn you that what you've just said is highly unorthodox and there's no serious evidence for it.'

'I told you, it was a joke.'

'*You* say it's a joke. Others may take it more seriously. You realize, it might even be *right*?'

'Hopper, calm down. You were telling me why the Planiturthians think space is expanding. So tell me this: if it's not because the universe is growing up, why *does* it expand?'

'Let's find out,' said the Space Hopper.

•

Vikki waited, but nothing much was happening. The nearby stars seemed to flicker a lot, but that was all.

'Hopper, is the VUE working?

'I don't see why not. We'll soon see. It's supposed to be running the universe backwards in time. To see where it came from.'

'Well, I can't see— *Hold it, that star just evaporated!* Look, it's turned into a cloud of gas!'

'Hydrogen and helium,' said the Space Hopper. 'Most of the matter in the universe used to be hydrogen and helium – the stars made the rest in their nuclear furnaces. When the universe runs backwards, all the heavier elements decompose back into hydrogen and helium. In fact, if I'm not mistaken . . .' he fiddled with the VUE and zoomed in on some of the nearby atoms '. . . yes, the mix is pretty close to one part helium, three parts hydrogen, as I expected. Now, let's see where *they* came from.'

'The helium is the big guys?'

'Yes, a helium atom contains two protons (the yellow blobs), two neutrons (green), and two electrons (pink), whereas a hydrogen atom is just one proton and one electron. And a proton is really a neutron with a different mix of quarks . . .'

'The atoms are breaking up! It's all going green!'

'That's right. We're back into the Era of Free Neutrons.'

'What's driving all this change?' Vikki asked.

'Well, in forward time the universe is expanding. So in backward time?'

'It has to be . . . contracting.' That made sense.

'Yes. And as it contracts, there's less and less space to hold the same amount of energy – so everything heats up.'

Vikki gasped. 'You're right! I can feel it!'

'I'm afraid some of the heat is seeping through the VUE's simulation boundaries. Don't worry, it will only get comfortably warm – I hope,' the Space Hopper added.

'What's the temperature now, then?'

'Ooohhh . . . no more than a billion degrees.'

Ulp! '*I hope you know what you're doing!* And space is contracting, you said? Shouldn't we get out before there isn't enough room for us to fit inside?'

'We're only "in" in Virtual Unreality,' said the Space Hopper, unruffled. 'We can exit the simulation before the universe gets *too* cramped. Or too hot. But we have to adopt an internal VUEpoint to see what's happening.'

They watched. Vikki imagined she could *feel* the universe shrinking around her. It was like painting yourself into a corner, except that the paint was nonspace.

'Now it's all going pink, Hopper, why's that?'

'The quarks have reassorted themselves into electrons. The temperature's risen to about two billion degrees, now, Vikki. And when it gets to three billion—'

FIZZZZZZZZZZZZZZZZZ!!!!!

'—a lot of vacuum will pull apart into electrons and anti-electrons. A billion times as many as there were just now.'

'I see. In backward time, they're unannihilating each other. And that means *we have only thirteen seconds left.*'

'Until what?'

'Until the End of the Universe! Rather, the Start of the Universe VUEd in backward time. I think it might be a good idea to slow down the VUE's clock.'

Backward seconds ticked away . . . backwardly. And, to their perceptions, more slowly.

'It's all gone sort of . . . *transparent.*'

'That's the neutrinos decoupling from the rest of matter. The temperature must be up to about ten billion degrees now. One second to go!'

'Nothing much going on at all, now, so far as I can tell. Are you *sure* the VUE is working? I think something's broken—'

'No, the temperature's up to thirty billion degrees, that's all – too hot for any atomic nuclei to stay together. It's just independent particles now. That's why we can't see much going on any more.'

'What's that wispy stuff, then?'

'Time. The VUE has decided to show us time as if it's something material. Watch the wisps shrink! A tenth of a second to go.'

FIZZZZZZZZZZZZZZZZZ!!!!!

'Ah. The unannihilation of the hadrons,' said the Space Hopper, as if he'd been expecting this. 'Yes. The heavy particles, like protons and neutrons and antiprotons and antineutrons, have multiplied a billionfold, just like the electrons did.'

'Whatever there was, there's a lot more of it now,' said Vikki, 'but crammed into a much smaller space.'

'More vacuum converted into particles and antiparticles, I'm afraid. It'll soon run out altogether.'

'What will?'

'The supply of vacuum.'

'Oh. I thought vacuum was nothing, anyway. How can nothing run out?'

'It's empty space, and the *space* is running out. Anyway, a quantum vacuum isn't really empty: it's a seething torrent of particles and antiparticles winking into existence and back out again, I've told you that. Look, the time-wisps are shrinking – time is *literally* running out, there's only a hundredth of a second left. I'll slow the clock some more or we'll miss it. Oh, and we're up to a hundred billion degrees, by the way, and that means we ought to—'

GLOMP

'—get out sharpish.'

Vikki was getting better at adapting to sudden shifts of VUEpoint, and the dizziness lasted only an instant. 'Is *that* the Planiturthian universe, Hopper?'

'Yes.'

'That soggy thing like a deflating football?'

'Yes. But a very hot football. Three hundred billion degrees and rising. And in a thousandth of a second . . . see how tiny the time-wisps have become? . . . it will—'

FLOOmp

'—disappear.'

They stared at empty metaspace. No particles, no vacuum, no temperature . . . and no time-wisps.

'That,' declared the Space Hopper proudly, 'is known as the *Big Bang*!'

'More of a Big Flop,' said Vikki. 'What an anticlimax!'

'That's because we saw how it looked *backwards,*' the Space Hopper pointed out. 'It's like reading a story starting from the end and working back to the beginning. Of *course* it's an anticlimax. But in forward time, it's like a sudden humongous explosion! Space, time, and matter all coming into being from absolutely nothing.'

'Why did you show me it backwards, then?'

'Because that's how the Planiturthians figured out what must have happened. Take an expanding universe and track it back in time, and you can see it must collapse. And that raises the temperature, and then all the matter comes to bits, and that leads to everything we've just seen. Tell the story in the right time-direction, and you have the story of the origins of the Planiturthian universe.'

It all sounded exceedingly far-fetched, even so. 'Is the expansion – the redshift – the only evidence for the Big Bang?'

'No, there are other things that corroborate it. You can even detect the Bang's echoes. No, I don't mean sound – "Bang" is a metaphor, and so is "echo". But electromagnetic radiation has been bouncing around the universe for twelve to fifteen billion years, ever since the Big Bang went off, and that radiation has been detected – and it looks right.'

'What *caused* it?'

'Sorry?'

'What caused the Big Bang?'

'There wasn't a cause. Couldn't be.'

'But – what happened *before* the Big Bang?'

'Vikki, there wasn't a "before"! Look at where the universe was in metaspace. What do you see?'

'Nothing. Just . . . metaspace.'

'Which is a Mathiversian fiction and doesn't really exist. Do you see any *space*?'

'No.'

'Does that bother you?'

'No. Space appears when the universe blows up.'

'Do you see any *matter*?'

'No.'

'Does *that* bother you?'

'No. Matter appears when the universe blows up, too.'

'Do you see any *time*? Any time-wisps hanging around out there?'

'No. They vanished along with the soggy football.'

'Does *that*—'

'You're telling me that *time* arises when the universe blows up?'

'Exactly. Without time, there can't be a "before". Since causes occur before their effects, there couldn't have been a cause, either.'

'But—'

'Of course, maybe there really is some sort of . . . extrinsic paraspace and paratime . . . and the Big Bang *did* have a cause,' the Space Hopper mollified her. 'Some Peoples think so. They even think that universes might bud off black hole babies that can *evolve*. So you never know. That's what makes physics such fun.'

•

Sixday 28th Noctember, 2099

Dear Diary,

The Space Hopper got so carried away showing me the ORIGIN of the Planiturthian universe that he completely forgot what we started talking about, namely, its SHAPE!

Until I reminded him.

Well, it turns out that the Planiturthians don't actually *know* what shape their own universe is! I told him that proves how ignorant they are, but he said that working out the shape of a universe *from the inside* is pretty tricky, especially if you can't go and visit places. All you can do is look in various directions, and see whatever you can see. Which may not be everything you'd *like* to see.

I said that we Flatlanders *know* that our world is a plane. He said we're right – as it happens – but that if we weren't, we might find it very difficult to discover we were wrong.

You know something? He's right, dammit. I hadn't thought about it before, but the problem is, we only ever see a limited part of our world. If it *looks* like part of a plane, we assume the whole world IS a plane. But for all we know, it might be the surface of a very big sphere, or a flat torus, or a hyperbolic plane, or a finite hyperbolic surface, or some kind of awful super-teapot with seventeen holes . . .

Basically, it's like trying to work out the geography of some distant country without ever being able to go there.

Except: you're allowed to *look*. And that opens up some fascinating possibilities.

Let me tell you what would happen if Flatland wasn't really a plane, but a flat torus. Remember Squarey Space? A flat torus is a square whose edges 'wrap round' and join seamlessly. The way Planiturthians always draw tori makes them LOOK curvy, but that's because they embed them in 3D space. If you don't bother about this, the tori can be flat. Which, you'll recall, just means that the angles of a triangle add up to 180 degrees, and so on. It doesn't mean you could *iron* them.

The easy way to think of a flat torus is to tile the plane with infinitely many copies of a square, and insist that whatever things you find in one copy, you will find it in exactly the same place in

ALL copies. Now, if Flatland were a SMALL flat torus, then you could look in certain directions and *see yourself in the distance.*

You'd immediately know you weren't on a plane, yes?

But what, Diary Dear, if you're on a *very big* flat torus – so big that light hasn't had time to go all the way across it? Well, you're then in much the situation that the Planiturthians are in right now.

The simplest possibility for the *spatial* shape of the Planiturthian universe is a 3-sphere, like the one in my nightmare, a few weeks ago (when I started painting a ball and got myself trapped *inside* the paint!). That fits the simplest Big Bang mathematics – space is a sort of 3D balloon that blew up very big. But there are other possibilities, depending on what the shape was right at the beginning and exactly how it grew. In the simplest of these, space is curved, but the curvature is the same everywhere – a nice, symmetric condition that fits rather nicely with the symmetry of natural laws. Of course that symmetry *could* be broken, but the easiest assumption is that it's *not*, OK? Well, that assumption leads to three possibilities: positive curvature, zero curvature, and negative curvature.

The 3-sphere has positive curvature, but so do other things – in particular various shapes that *tile* a 3-sphere, with suitable 'wrap-around' rules for going across edges. Do that with half a sphere, by the way, and you get the confounded Projective Plain – again. There's a 3D flat torus – a *cube* with opposite faces wrapped round – and that has zero curvature. And in the negative curvature department there's 3D hyperbolic space, plus whatever you can extract from it by using a shape that tiles it. With wrap-around rules as well, of course, or else your universe would have an edge – and that's bad because the laws of nature would be different at a genuine edge: you wouldn't be able to *go beyond it*, and that's contrary to symmetry, you see.

Anyhow, in any of these finite universes with wrap-around rules, you're in much the same situation as a Flatlander in a flat torus. If the universe were so small that light could cross it very quickly, then in principle you could look in various directions and see yourself. By mapping all the copies of yourself that you could see, you could work out whether the universe had positive, negative, or zero curvature, and what shape it was.

But in a *BIG* universe . . . well, let me show you how it goes on the flat torus. The region of space that's visible from a given point forms a circular disc, and as time passes that disc grows. When it grows to roughly the same size as the whole universe, it starts to OVERLAP ITSELF. At that point – and not before, Diary Dear, because of the speed of light, I'm sure you can see that – it becomes possible to observe the same object in more than one direction. Any object in the overlap will do, but the others won't, of course. The first objects you can see that way will be sitting at the points where the boundary of that disc overlaps itself.

As time passes . . .

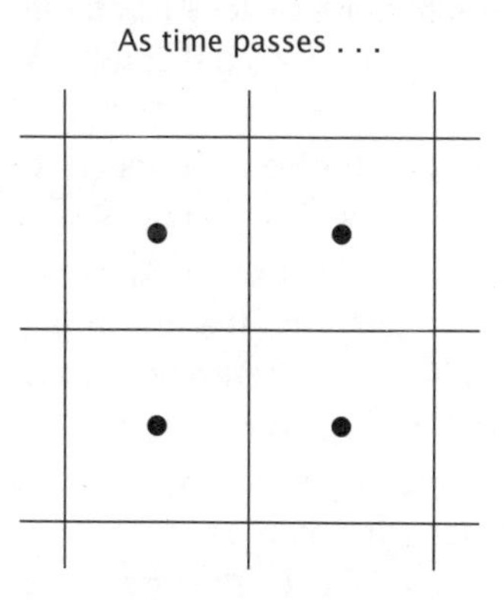

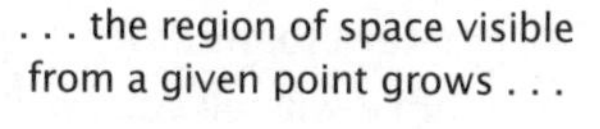

. . . the region of space visible from a given point grows . . .

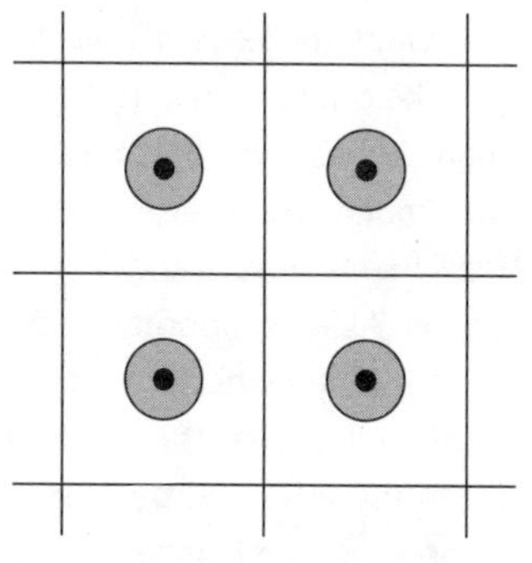

. . . until it touches itself . . .

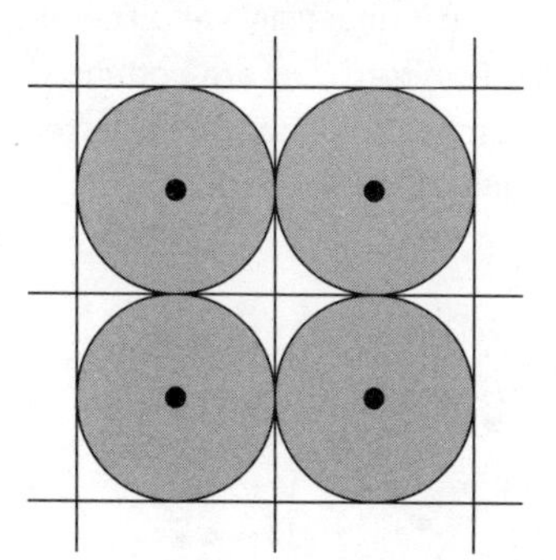

. . . and then overlaps itself

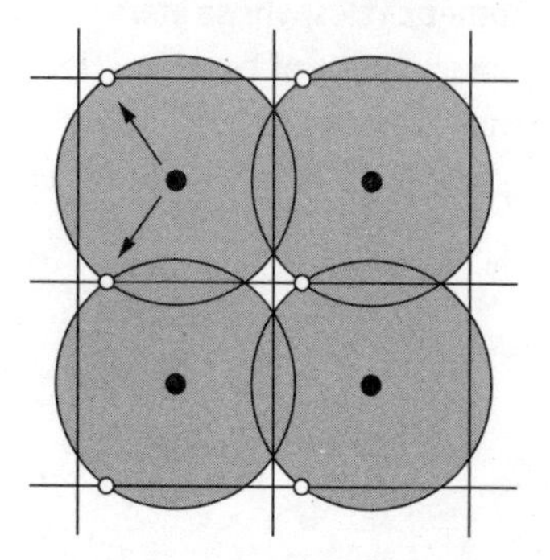

From then on, some regions of space can be seen in several different directions

Now, the Planiturthians start with a few advantages. They reckon they know roughly how old their universe is, and roughly how big (because they know how fast it expanded from the Big Bang). And they've calculated that their light sphere, if it exists, ought to have started to overlap itself by now.

So the Planiturthians are planning to hunt the overlap down. Since their space is 3D, the disc has to be replaced by a solid 3D ball. So the places where it first overlaps itself form big circles in the sky . . .

Why circles? Because, Diary Dear, a sphere intersects another sphere in a circle. Even if the 'other' sphere is just a different bit of the first sphere that's been wrapped around and overlaps!

To return to my story: the pattern of those circles – if you can find them – will tell you what shape your tile is, and that will tell you what shape your *universe* is.

The only problem is that what the Planiturthians see repeated many times over on those special circles is either empty space (not much use, Diary of Mine, since *all* regions of empty space look identical) or randomly scattered stars. And the rest of the sky is ALSO randomly scattered stars, *and* the Planiturthians have no idea where to look for those circles. So picking out the right circles from the random speckling of the night sky is next to impossible. The best they can do is trial and error – mostly error. Pick some circle – which may or may not be the right one – and record the positions of its stars. Then search the sky for other circles whose stars seem to match it unusually well. There's an awful lot of circles to try. But the Planiturthians are nothing if not persistent, and their answer to that is to build faster computers and just crunch circles until they find a match.

Or not.

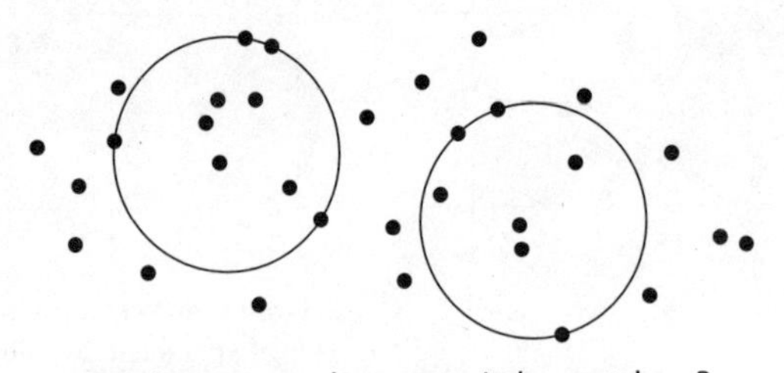

Do the stars on these two circles match up?

Intelligent it isn't. Except that there doesn't seem to be a cleverer way to go about it.

You know, I think that when I get back to Flatland I'll try to find out what shape *that* really is.

Bother, now I've reminded myself. I'm homesick, Diary. The Mathiverse is tremendous fun, and the Space Hopper is a competent, if sometimes irritating, guide – despite being utterly weird. But in the early hours of the morning, when I find myself awake and can't get back to sleep, I miss my home and my family more than I dare admit.

And I *hope* they're missing me, and I'm worried that they may very well be doing just that. And I'm even more worried that they're not.

I hope they found the Crisp Moose card I sent them.

NO-BRANES AND P-BRANES

As usual, the opportunity to bring up the topic of a return home never arose. The Space Hopper was far too enthusiastic about whatever bit of the Mathiverse he wanted to show off next, and Vikki was too kind-hearted to disappoint the poor creature. And its ∪ was kind of cute, in a ghastly sort of way.

They were discussing the state of Planiturthian physics – the nature of space, time, and matter.

'Of course, the Planiturthians weren't content to leave it at that,' the Space Hopper observed. 'And you can see why.'

'I don't see why at all,' said Vikki. 'The Planiturthians had worked their way towards one amazingly effective and accurate theory of the large-scale structure of space, time, and matter – and a different, even more amazingly effective and accurate theory of the small-scale structure of space, time, and matter. So between them, those two theories covered everything.'

The Space Hopper's ∪ widened. 'Actually, that's the one thing they didn't cover.'

'What?'

'Everything!'

'You mean – they didn't cover the medium scale?'

'No, I mean they didn't cover *everything.* Not all in one go. What the Planiturthians wanted was a Theory of Everything. *One* theory. Not two. One set of laws to explain the large scale and the small.'

'I must be missing something here,' said Vikki, deeply perplexed. 'Why can't they have one theory with two options? *These* laws for the large scale, but *those* laws for the small scale. Like groceries – one price for a small order, but a different price for a big one. I don't see the problem.'

'Well . . . I guess they just felt that kind of conditional law wasn't elegant enough.'

'Oh,' said Vikki. 'They wanted it *elegant*, too? Why not just stick to something that worked?'

The Space Hopper bobbed uncomfortably. 'Because it didn't, not entirely. I know, it sounds all right in theory: one set of laws for the large scale, a universe-sized law; another set of laws for the small scale, a subatomic-sized law. And never the twain shall meet. But then came the Big Bang, as you already know. And in the Big Bang—'

'What ends up as a universe-sized object starts out as a subatomic-sized one! And it grows continuously from small to large . . . so there's no obvious place to swap the laws.'

'Correct. And in some respects it was even worse than that. You see, the large-scale laws had implications for the small scale, even after the Big Bang was long past. That's because the large-scale laws indicated that there was a "force of gravity" that acted on the small scale. Between any two particles.'

'I thought General Relativity explained the force of gravity away by reinterpreting it as the curvature of spacetime.'

'Yes, but that really works only on the universe-sized scale. It doesn't fully explain why the solar system behaves as if each body attracted all the others. Sure, you can come close . . . but what would really sort everything out would be a gravitational particle. A graviton. Just as a photon is a particle of light, so a graviton should be a particle of gravity.'

'Hold it. Gravity is a force. Light isn't.'

'On the face of it, that's true. But it's not that simple. And there are other types of particle for which the analogy works better.'

'Oh.'

'And just as the large-scale laws had implications for the small scale, so the small-scale laws had implications for the large scale. They meant that in a sense the large-scale universe didn't really exist at all. Because the mathematics of General Relativity assumes that the universe is an infinitely divisible spacetime continuum, whereas the quantum theory says that space, time, and matter can be subdivided only so far before they become indivisible. You can subdivide them a long way, I grant, but not for ever.'

'That's something of a hair-splitting argument.'

'Hair-not-splitting, really. But you're right, it wouldn't have mattered so much if the Planiturthians hadn't been so set on finding a Final Theory that unified the whole shebang in one fell swoop. Wrapped up the entire ball-game, all at—'

'Hopper, I think you're overdoing it. I get the picture. Anyway, why would anyone want to do that?'

'Well, the Planiturthians had all sorts of rational reasons, but when it comes down to it, I think it was because Planiturthian science had grown from a monotheistic religion. Most of their science was the product of a culture whose religious beliefs attributed their entire universe to a single act of creation by a single deity. Whether or not the scientists themselves were religious, this cultural trait led them to seek a single explanation for *everything*. The search for One Final Theory is psychologically very close to the search for One True God – though the Planiturthian scientists wouldn't have agreed with that view at all. They didn't like religious explanations. None the less, they adopted a very fundamentalist attitude – they used that word a lot, "fundamental". They seemed to think that their mathematical equations – if they ever found them – would be How the Universe Really Works. Which is a breathtaking piece of arrogance, coming from a bunch of creatures that have experienced only one tiny planet, and then only for a short period of time and on a small range of scales.

'But maybe that's being unfair. A lot of them understood that their hoped-for theory would be fundamental only in a metaphorical sense – that *in principle* it might be possible to base most of their other theories on it. What they expected to be done in practice was much less ambitious. And I think some of them even realized that what they would get would be a description of how the universe seems to work, not the actual rules by which it runs. Because – well, because there might not be any such rules. A *rule* is a very Peoplish concept, and even more so a *law*. Those are methods Peoples use to run their own society. So it seems to me that they had a mental picture of their universe that was modelled on their own social interactions. The amazing thing is that it worked pretty well.'

•

Sevenday 29 Noctember 2099

We're getting into very difficult intellectual territory here, Diary Dear, and not just philosophically. This stuff is right out on the frontiers of the Mathiverse, where new concepts are still popping into existence . . . and the physics is pretty hairy too. So I'm going to make some notes for you.

The most important item of background information, the one that sets the problem up, so to speak, is that the Planiturthian universe seems to be governed by exactly four different kinds of force. What's a force? Something that lets matter affect other matter. How does it work? Forces are produced by *fields*, which in turn are associated with bits of matter, and other bits of matter respond to those fields. For instance, a magnet is surrounded by a magnetic field, and any matter that responds to magnetism is affected by whatever magnetic field it finds itself in.

Got that? Good.

What are the four forces? Here goes:

- *Gravity*. This is a very weak force – it takes the entire mass of Planiturth to generate enough gravity to hold Planiturthians on the ground, and even then they can jump to roughly their own height (with training, effort, and skill). In compensation, though, gravity is a long-range force – it acts over extraordinarily large distances. Right across the universe, in fact – it's what holds their universe together.

- *Electromagnetism*. Originally the Planiturthians viewed this as two different things: magnetism, which made needles point north and lined up iron filings in patterns; and electricity, which gave them shocks when it built up on metal objects, and powered their lights and televisions. Both magnetism and electricity have fields associated with them, which you can measure with the right instruments. After the work of Jamesclerkmaxwell, who invented a single mathematical framework for both the magnetic and electric fields, the Planiturthians recognized that they'd merely been seeing two different aspects of the same thing. Electromagnets can turn electricity into magnetism, and dynamos can turn magnetism into electricity. This was the first *unification* of apparently distinct physical fields and the associated forces. Electromagnetism is intrinsically stronger than gravity, but it comes in two kinds – positive and negative. Because those tend to cancel out, the dominant large-scale force in the universe is gravity, not electromagnetism. This is why General Relativity works pretty well on cosmological scales.

- *Strong nuclear force.* You don't come across this directly in everyday life, Diary! However, without it there wouldn't BE any everyday life, because it's what keeps protons and neutrons together in the atomic nucleus. No strong nuclear force, no nuclei; no nuclei, no atoms; no atoms, no molecules; no molecules, no anything. The strong nuclear force really *is* strong, which is why nuclei are so hard to break up, but it acts only at short range – about a quadrillionth of a metre.

  More accurately, the strong nuclear force acts between quarks, which are the fundamental (that word again!) constituents of protons and neutrons. Quarks are complicated beasties, and, as you know, a proton or a neutron contains a bunch of them . . . so the strong nuclear force is a rather messy thing to work with.

- *Weak nuclear force.* A lot of particles – the technical terms is 'leptons' – don't 'feel' the strong nuclear force at all. But they do feel the weak nuclear force, surprising as this may sound. Basically, some particles are 'immune' to the strong nuclear force – the Force isn't *with* them, OK? The weak nuclear force is a lot weaker than the strong nuclear force (surprised?) and has about a hundredth of its range.

OK, Diary Dear: now you have the *ingredients*. The problem is to make sense of it all!

That's hard.

•

'I still don't fully follow this business of fields and forces,' said Vikki in a troubled tone. 'It seems to mean that in some mysterious way a particle can affect another, possibly quite distant, particle, without anything passing between them.'

'In a classical picture, *and* in a relativistic one, that's true, and it's not just a philosophical difficulty. It's one reason why Planiturthian physicists found it necessary to try to marry gravitation and quantum theory. In a quantum picture – well, let's take a look.'

The VUEfield zoomed in on their surroundings, homing in on a fuzzy pink ball. 'Look, an electron!' said the Space Hopper. 'Listen closely, and you'll hear it singing.'

'Electrons don't sing!'

'VUE-enhanced electrons do. For heaven's sake, Vikki, we *spoke* to the Charge on the Electron not long ago. Don't balk at a bit of harmless singing! At least they're not dancing too, like the Space Girls did. Turn up the volume, and you'll hear it.'

Vikki did as she was bid, and sure enough, a high-pitched voice was singing a curious, repetitive little song:

'I can't *get* no / more mo*men*tum / got fixed *en*ergy / can't in*vent* 'em . . .'

'What's all *that* about?'

'It's singing about conservation laws. Which is nice.'

'No, I mean, how come it's singing at all?'

'It's how the VUE represents a quantum field. Electrons don't really sing—'

'That's a relief!'

'—they hum.'

The singing went on for several minutes, and then a second, lower-pitched voice started to join in, in a weird descant:

'. . . *take* you down / 'cos I'm *go*ing to / Yang-Mills *Gauge* Fields / Nothing *is* real . . .'

A second pink ball came into VUE, closing in on the first one. 'This should be interesting,' said the Space Hopper.

'They're on a collision course!' cried Vikki.

'Maybe. They'll come close, certainly.'

As the two particles approached each other, their songs became strangely intermingled, and odd snatches of other songs began to intrude:

'I can't *get* down / 'cos I *really really want* / got nothing . . .'

'. . . *take* you no more / fixed *en*ergy Fields / did it *my* way / *is* real . . .'

'They're interacting,' said the Space Hopper.

'Sounds more like *over*acting,' Vikki commented. 'They're going to *hit*— Hey! What's that?' But as the Space Hopper started to answer, Vikki interrupted: 'Hey – they *bounced!*'

'Ho-o-o-*on*ky / -tonk *lep*tons! / Gimme, gimme, *gim*me / the honky-quark-gluon-*plas*ma . . .'

'You ain't nuthin' but a hadron / nucleifyin' all the time / You ain't . . .'

'And their songs have changed.'

'Yes,' confirmed the Space Hopper. 'The collision has altered their quantum states – energy, momentum . . . *could* have been a change in charge, too, but not this time.'

'Why?'

'You noticed something when they came close to colliding, didn't you?'

'Yes, it was a sort of red squiggle.'

'That was a photon. They exchanged it.'

'Why?'

'Well, in the quantum world, energy and momentum and charge and suchlike are all quantized – they come in whole-number multiples of tiny, irreducible amounts. When particles interact, their quantum fields affect each other, creating forces between them, and the result is to *change* their energy, momentum, charge, whatever. That's how the electrons bounced. *But*, for the reasons I've just given, those changes come in whole-number chunks. A tiny whole-number bit of momentum, say, gets transferred from one electron to the other. Something has to *carry* that chunk of momentum – take it away from one and give it to the other. While the momentum is in transit between the two electrons, it in effect forms – or more properly, can be represented as – a tiny particle of its own. In this case, a photon.

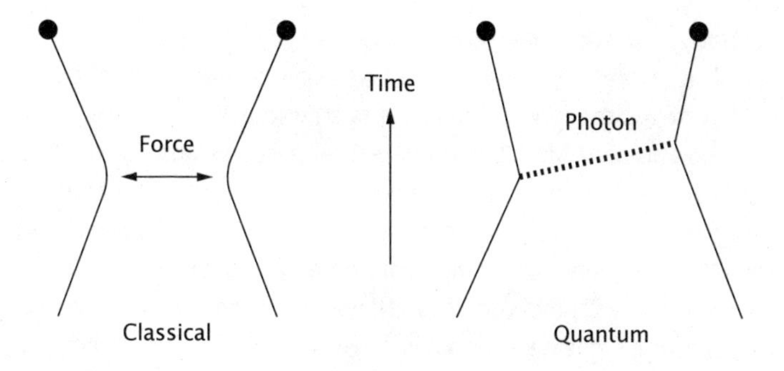

'*All* quantum force interactions are carried by particles in this kind of way', the Space Hopper went on. 'Electromagnetic forces are carried by photons, strong nuclear forces by eight different gluons, and weak nuclear forces by the W^+ boson, the W^- boson, and the Z boson.'

Vikki mused on these remarks. 'I thought I saw a purple squiggle passing from one electron to the other, as well as a red one. Was that a gluon or a doubleyouplusboson or something?'

'A something,' said the Space Hopper. 'That was the infamous graviton, and you were extremely fortunate to see it. The graviton is the carrier particle for the force of gravity – assuming that any such particle exists, which for narrative purposes the VUE just did. Any unification of Relativity and quantum theory has to bring the gravitational force into line with the other three – so, in particular, it has to introduce a particle to carry the gravitational force.'

The Space Hopper warmed to his theme. 'But that's not so easy. The scheme that works for the other three forces – known as the Standard Model – is pretty complicated. You can't bring in a new particle without upsetting all the delicate logical implications of the mathematics behind the Standard Model. The newcomer has to be introduced with care, forethought, and the very best taste. Otherwise all it will do is disrupt the logical harmony of the Standard Model, with disastrous consequences.

'And one of the main features of the Standard Model – one that pretty much qualifies it as a Geometry in the Felixkleinian sense – is symmetry.'

•

Wunday 30 Noctember 2099

Well, now we're getting to the heart of the matter, Diary Dear. What drives the whole enterprise is a curious kind of symmetry of the Standard Model of fundamental particles and forces. I've told you already that the deep nature of a Geometry – that is, of a Space or a Spacetime – is best captured by its symmetries. If you're a mathematician, the symmetries characterize the geometry; if you're a physicist, they tell you what the possible laws of nature are – because the laws have to be *invariant* under the symmetries, OK? The laws must respect the symmetry, and not bash it about.

The most obvious symmetries of the Planiturthian universe are those of Space and Time. Every region of space behaves, in principle, like any other; ditto for instants of time. I don't mean that every bit of space looks the same as every other – you might have an atom of carbon at one point of space and vacuum at another. But in principle it could have been the other way round: there aren't regions of space that *must* have carbon atoms in them, or *can't*.

Those symmetries imply that the laws of nature must be the same at all places and at all times. But they don't give much more of a clue as to what those laws ought to be. A more interesting restriction comes from *time-reversal.* If you run the universe backwards in time, it obeys the same laws that it did in forward time. That symmetry is more puzzling – especially since the Planiturthian universe seems to run in just ONE time direction. But again, the point is not that some particular realization of the universe possesses time-reversal symmetry; just that whichever behaviour you encounter, it could equally well have run backwards in time without the *LAWS* changing.

However, there are other, less obvious symmetries on the quantum level. For instance, you can 'reflect' an electron – not in an ordinary spatial mirror, but in a 'charge mirror'. By which I mean you can change its charge from negative to positive. When you do that, the electron becomes a different particle: a POSITRON. However, the laws that govern positrons are just like those that govern electrons, provided you 'reflect' all charges.

Well, nearly. I lied, OK?

In fact, there are slight differences: the charge-reflection symmetry is 'broken' in the actual Planiturthian universe. Some particle interactions involving positrons look slightly different from the corresponding interactions involving electrons. An attractive explanation for this breaking of symmetry is that at the time of the Big Bang, the nascent universe possessed exact charge-reflection symmetry, but then – as the universe expanded – that symmetry became broken. This idea has another advantage: it suggests that at the time of the Big Bang, ALL four forces of nature were unified into a single force. All the different carrier particles for those forces were in effect the same kind of particle – and the differences we now see arose when that original symmetry broke.

So this setup provides a scheme for unifying all the forces of nature – a plausible route to a Theory of Everything. All you have to do is set up a really symmetric version of physics, appropriate to the time of the Big Bang, and then break the symmetry in the right way.

Not easy. But symmetry, Diary Dear, is the key.

•

'It's like politics,' said the Space Hopper. 'All a question of spin.'

'Spin?'

'In certain respects, quantum particles behave *as if* they are spinning around some axis. Like a top.'

'You mean, if we used the VUE to look closely at an electron—'

'Let's do that and find out. But I warn you, what we'll see is really too classical an image. The spin of a quantum particle isn't really *movement.* At least, if it is, it's not movement in *space.* Think of it as some kind of intrinsic excitation – a "state of mind", if you wish. But, for now, let's program the VUE to realize spin as . . . spin.'

No sooner said than acted upon: a fuzzy red ball rotated gently before their eyes. The VUE caused a thin blue line, the axis of rotation, to become visible, and decorated it with an arrow to indicate the direction of rotation.

'That's a photon,' said the Space Hopper. 'Its spin is 1 quantum unit. Now this guy over *here* . . .' he gestured towards a fuzzy brown ball '. . . is an electron, with spin ½. Which, by the way, is the minimum size that quantum spin can be. Except zero, of course.'

'Shouldn't the minimal unit be spin 1?' Vikki objected.

'Well,' said the Space Hopper, 'spin ½ is more traditional.'

'What's tradition got to do with it?'

'When there are Planiturthians around, lots. And there are . . . other reasons. For that choice of units.'

'Such as?'

'Particles with whole-number spins, like the photon, behave just like classical spinning particles in one important respect: what happens to their spins when you twirl their axes in space. Keep a close eye on that photon, and I'll show you.'

Vikki aligned herself with the photon's axis, so that the blue arrow showing the direction of spin ran clockwise. The Space Hopper grabbed the axis of the photon and began to move it. 'Hard work, this, it's like trying to turn a gyroscope,' he muttered. 'All that angular momentum – wears a chap out.' Slowly, the photon's axis turned; the photon continued spinning, without visible interruption or change. After a lot of shoving and swearing, the photon's axis had turned through a complete circle. The arrow continued to point clockwise, a fact that the Space Hopper pointed out with considerable excitement.

'Don't be so over-dramatic, Hopper!' protested Vikki. 'That's exactly what you'd expect!'

'Ah, but this is the quantum world, where the wise person expects the unexpected. Let's try it with an electron – which has spin ½ – instead.'

Vikki lined herself up with the particle's axis, as before, and again the blue arrow showing the direction of spin ran clockwise. The Space Hopper moved the axis in a circle, still complaining under his breath. After a lot of shoving, the electron's axis had turned through a complete circle.

The arrow now pointed anticlockwise.

'But – that's mad,' said Vikki.

'Welcome to the quantum world.'

'You must have turned it through 180 degrees by mistake.'

'No, it was a full circle, 360 degrees, I guarantee. How far must I turn the axis to get the direction of spin back to clockwise, Vikki?'

'Well – if a 360-degree turn of the axis in space reverses the spin, I suppose that a second 360-degree turn of the axis in space would reverse it again. That means that after a 720-degree turn of the axis in space, the direction of spin should come back to where it was.'

They tried it.

It worked.

'This is very puzzling,' said Vikki.

'It illustrates one of the ways in which quantum spin is *not* rotation about a spatial axis,' said the Space Hopper. 'That behaviour is a consequence of the mathematics of quantum spin. And it means that quantum particles come in two very different kinds. They're called *bosons* and *fermions*. Bosons have whole-number spin – an even number of basic spin units of ½ – but fermions have an odd number of basic spin units of ½. The photon, with spin 1 – two units of ½ – is a boson. The electron, with spin ½, is a fermion. And if the much-sought graviton, the "missing link" between quantum theory and Relativity, exists, it should have spin 2 – making it a boson – and (as it happens) mass 0.

'Anyway, bosons behave like classical particles when their spin axis is rotated. Fermions don't. There are all sorts of other differences, too.'

Vikki leaned back to digest all this. 'It's complicated, Space Hopper.'

'Sure is. I'm skimming the surface of a detailed – and sophisticated – piece of geometry, OK? Don't expect to get it all, I haven't *told* you it all. The big point here is that there are these two very different kinds of particle, with different kinds of spin, as unlike each other as odd and even.'

'Got that.'

'So of course, Planiturthian physicists got used to thinking of them as completely different things.'

'Fair enough.'

'Until one day they discovered that bosons and fermions could be viewed as two different aspects of the same thing. There was a symmetry of the universe – or, at least, of a mathematical scheme to represent the universe – that could turn bosons into fermions, and vice versa. They called this *supersymmetry*. It implied that associated with every particle there should be a corresponding superparticle – and if the particle is a boson, then its superparticle is a fermion, and vice versa. It was a huge surprise.'

'I'll bet,' said Vikki. 'But how can a symmetry of the universe change an even number of spin units into an odd number? Won't that change the laws?'

'Oddly enough, no – one reason why this was an exciting discovery. It's because of the way quantum spin works.'

'So the VUE ought to be able to start with a photon, apply a supersymmetry, and give us an electron?'

'It's not as easy as that. There's no reason why the superparticle corresponding to a known particle has to be another *known* particle. In fact, the superparticle corresponding to a photon isn't – it's believed to be a hypothetical particle called a "photino". Let me show you how supersymmetry turns a photon into a photino.'

Vikki watched as the Space Hopper captured a fuzzy red photon and lined it up for a supersymmetry transformation. 'Here we go!' he yelled. The fuzzy ball promptly turned brown, and the Space Hopper beamed with pride.

'Um,' said Vikki, in a bored voice, 'was that it?'

'What? Not impressed by my marvellous – oh, you weren't using the full power of the VUE! Vikki, you really must pay attention to *everything* that the VUE can show you. Let's try it again, but first – activate *this* VUE-extension, OK?'

The Space Hopper recaptured the errant photino. 'I'll turn it back into a photon. Tell me what you see as I'm doing that.'

'Well, it starts out as a brown ball – no, wait, hold it. It's not a ball and it's not brown. What have you done now?'

'I've set the VUE to show you the hidden dimensions of superspace.'

'Oh. Well, it *is* a ball, but it's a much higher-dimensional ball, and it's all sorts of colours. But there's a sort of brown stripe—'

'That's the photino state,' said the Space Hopper helpfully.

'—and another red stripe—'

'The photon state.'

'—but they're not really stripes, they're fuzzy balls of lower dimension . . . sort of, it's hard to describe in words.'

'Count the dimensions of spacetime that are present in the image, Vikki.'

'Uh – one, two, three, four, five – *five*?'

'Keep going.'

'Er – five, six, seven, eight. *Eight*?'

'Yes. Four dimensions of spacetime, and four more, called superspace. According to the theory of supersymmetry, every particle in ordinary 4D spacetime carries along with it a kind of attached ghost in a second, nonphysical 4D space – its superpartner. Well, really I should say that superspace is a *physical* but not a *spatial* 4D space, I guess. Tell me, now: which component is the brown stripe in?'

'Um – ordinary space.'

'That's right. So we're seeing a photino in ordinary space, but there's also a lot of attached ghostly gadgetry that we don't normally observe, which extends into four further dimensions – inside which there is what?'

'A red stripe. A ghost photon? In superspace, not real space?'

'Exactly. Now, all we have to do is *rotate* the whole picture so that real space and superspace get swapped – see, in 8D it makes perfect sense – and once we've done that, what do we get?'

'A red photon in real space and a brown photino in superspace. The ghost has become real and the real one has turned into a ghost!'

'Elegantly put,' said the Space Hopper. 'And *that* is supersymmetry.'

•

Twoday 31 Noctember 2099

Deep stuff, Diary Dear. The untamed frontiers of the Mathiverse are looming before me. And back home (sniff) it's New Year's Eve. Everyone will still be eating the leftovers from Crisp Moose Day, and waiting to welcome in a new year and a new century . . .

I'm getting maudlin. Back to the untamed frontiers.

Ingredients: four forces of nature. Three unified by advances in quantum theory; the fourth – good old gravity, still holding out.

Needed: a new particle, the graviton, to carry that force. Spin 2, mass 0: impossible in conventional quantum theory.

Implication: reformulate quantum theory AND Relativity. And you can't do it with point particles.

Required structure: hierarchies of particles, each with its own peculiar properties, somehow all part of one unified picture.

Hint: *SUPERSYMMETRY*. There's a new kind of extended space out there, superspace. Particles spill over into that, as well as into ordinary space.

So what, Diary Darling, do you think we are missing? What would tie the whole picture together?

That's right.

String.

•

'Superstring, actually,' said the Space Hopper. 'If you want to bring gravity into the scheme of things, you can't keep the Standard Model exactly as it is. You have to be prepared to modify it a little. Hah! I say "a little", but that "little" includes a complete rethink of the structure of spacetime. Not just a supersymmetric add-on. Something more significant than that.'

It seemed to Vikki that this was going Over The Top. 'Why does spacetime have to change, Hopper?'

'Because when you try to quantize gravity, it turns out that particles can't actually *remain* particles. That just doesn't work – pointlike objects just don't fit all the requirements. So they have to be replaced by something else.'

'What?'

'Like I said: superstrings.'

'Yes, I know that's what you said, but it made no sense then and it doesn't make any more now!'

The Space Hopper gave the criticism due consideration. 'Very well,' he said. 'A particle is a point – or, at least, it looks like a point, with no internal structure. It's a pointlike object, yes?'

'Sure.'

'Moving up in dimensions, you could replace a pointlike particle by a curve – a string. Come to think of it, Vikki, *you're* a string. Like all Flatland women. Geometrically, you're a line segment.'

'Do quantum strings have two ends, like I do?'

'Some do.'

A constant worry of Flatland women surfaced. 'Do you think my endpoints make my bum look flat?'

'I'm not a good judge of Flatland womanhood,' the Space Hopper said tactfully. 'But it seems to me that – er – *quantitatively* speaking, you have the exact correct number of endpoints. One at each end, which is ideal.'

Vikki decided to be flattered. 'That's very kind of you. So what happens to quantum strings without ends?'

'They loop round and form a closed loop. A topological circle.'

'Ah.'

'That's most of them, actually.'

'Tidier that way. Otherwise the ends would flap around.'

'They can, in some string theories. Anyway, curves are intrinsically one-dimensional. The next step up from curves leads to two-dimensional surfaces – membranes. Perhaps with exotic topologies, like Moobius, or the Projective Plain, or a doughnut.'

'Don't remind me about Moobius! That *cow*!'

'Well, she *was* a cow,' the Space Hopper pointed out.

'That doesn't justify her behaving like one!'

'Beyond membranes,' the Space Hopper said, diplomatically changing the subject, 'are three-dimensional analogues of surfaces, which the Planiturthian physicists insist on calling 3-branes. And then come 4-branes, 5-branes and so on. They also insist on using the symbol p for an arbitrary number of dimensions, rather than the customary N . . . and I think I know why. Guess what you get in dimension p?'

'Oh no! p-branes?'

'Absolutely.'

'So a surface is a 2-brane?'

'Yes.'

'And a string is a 1-brane?'

'Naturally.'

'Which makes an ordinary particle, with dimension zero, into a no-brane?'

'I had a feeling that's where you were headed', said the Space Hopper.

•

'You said "*super*strings",' Vikki remembered. 'You've only explained strings.'

'Well, you remember that particles – no-branes in your terminology – are accompanied by ghost extensions into superspace?'

'Yes. You mean superstrings are strings that extend into superspace as well as ordinary space?'

'Correct. It's quantum string, it can do that.'

'Quantum things can do anything you want, it seems to me. Where do you find these superstrings? At a superstationers?'

'In yet another extension of ordinary space, Vikki. Good job I updated the VUE's HyperZoom facility. Let me just check it's been installed . . . good. Follow me!' And down they went, into the smallest scales of the Planiturthian universe – into a world where an electron was too huge to contemplate, where its component quarks glistened like tiny jewelled specks.

'Right now we're at the limits of the VUE's *spatial* resolution,' said the Space Hopper. 'Down at the Planck length, where quantum effects make space so fuzzy that it's not even clear that space really exists at all.'

'It's frothy,' said Vikki, 'not fuzzy.'

'That's quantum foam – particles springing into and out of existence, creating space and time along with themselves. Try some, it's quite tasty.'

'No thanks!'

'Go on. It's a bit like chicken.'

'I'm not hungry.'

'Suit yourself. See those bright specks?'

'Yes. They're quarks, aren't they?'

'Among other things, yes. What do they look like on this scale?'

'Dots. Points. No-branes'

'Yes. And yet, those points have to support all sorts of quantum states. Spin, charge . . . would there be *room* for such diversity on a no-brane?'

That thought hadn't occurred to Vikki before. 'In what sense?' she asked, cagily.

'All will shortly become clear. I'll use the HyperZoom to improve the resolution in any other dimensions that may be around—'

'Like superspace?'

'Yes, but there might be others – there's more to a particle than its spin, you know. Ah, yes . . . coming into focus now . . . Look at that!'

'It's – hey, that quark's not a point at all! It's a tiny loop!'

'Yes.'

'And . . . isn't it *vibrating*?'

'A little bit, maybe. This one's in its ground state, lowest energy. But if I give it a—'

TWANGGGGGGGG!

'—then something really interesting happens. Zoom out for a moment, and you'll see.'

'Oh. It's changed colour!'

'Yes. That's the VUE's way of showing us that it's actually become a *different particle* altogether. A much more massive relative of a standard quark, as it happens. Doesn't have a name. Anyway, the main thing here is that anything from a 1-brane up can vibrate in all sorts of different ways, whereas a no-brane can't. That means that a 1-brane or a 2-brane or a 3-brane or a zillion-brane can easily support all sorts of different quantum states. Whereas for your traditional point particle the states are just some kind of unexplained add-on.

'So here's the idea. On really small scales, spacetime isn't 4D at all. What looks like a tiny point in 4D spacetime is really something else – some kind of p-braned topological hypersurface in a higher-dimensional space.'

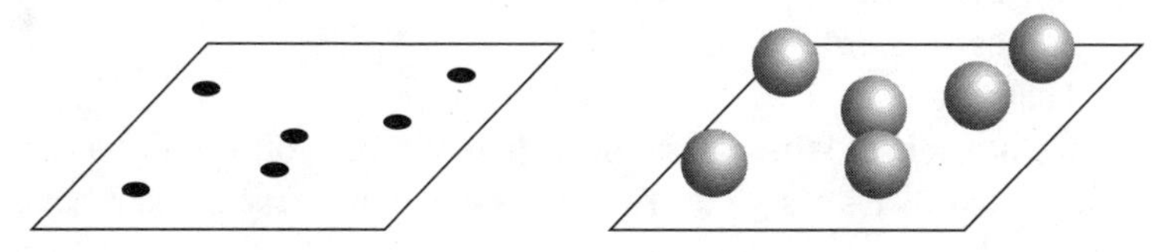

Every point in spacetime . . . . . . is really a multidimensional hypersurface

'How many? Dimensions, I mean? Equal to p?'

'No, could be bigger. For a while 7D and 22D were hot favourites, making the revised spacetime come to 11D or 26D. Right now, though, the consensus is settling on 6D for the extra dimensions – a total of 10D altogether.'

'Spacetime is really ten-dimensional?'

'That's what every well-informed Planiturthian thinks right now.'.

'You mean there's a choice?'

The Space Hopper laughed. 'Oh, yes! A huge choice. The range of *n*-dimensional spaces is inexhaustible! The problem, Vikki dear, is cutting that number down enough to be able to fix the exact one. Not just the value of *n* – the number of dimensions – but the topological shape of the corresponding space.'

'Put that way, I see the difficulty. So how *do* they sort out all those possibilities?'

'Still working on it. But they're making progress. Turned out that the original 11D model – supergravity, it was called – had a fatal flaw. It couldn't encompass the broken symmetry of the weak nuclear force. A People called Edwitten pointed that out in 1984, by their calendar. The way to deal with that was to reduce the dimensionality to 10D.

'Now, when you do quantum theoretic calculations, as often as not you get stupid answers.'

'No I don't.'

'Why not?'

'I don't do quantum theoretic calculations at all.'

'Ah. When *anyone* who does quantum theoretic calculations *does* do quantum theoretic calculations, as often as not they get stupid answers.'

'Easy to make mistakes?'

'No, even when they get the calculations right, they still get stupid answers. Infinity, usually.'

'That's bad?'

'It's nonsense. When a physical theory gives you infinity as an answer, it's always a sign of trouble. The real universe doesn't have infinities.'

'Hang on, Hopper – we went to infinity in the Projective Plain.'

'Yes, but that wasn't a real universe, OK?'

'Only an idealization, that's true.'

'And in a sense Infinityville wasn't a real infinity, either. But you're distracting me. There is a way to get round these quantum infinities, and it's called renormalization. Doesn't matter exactly what it is – just take my word that it's not an easy trick to pull off. So most potential superstring theories don't work because they're not renormalizable. It turned out that for a 10D superstring theory to be renormalizable it has to be one of five competing possibilities.

These are known by their symmetries. One is called SO(32), another is $E_8 \times E_8$, and they both have an extra 16 dimensions of 'internal' states, as well as the 10D of the string itself. See? 26D in all. And the other three are known as Type I, Type IIA, and Type IIB.'

'Imaginative terminology.'

'Very. The fascinating thing – a strong hint that such theories could unify gravity and quantum theory, the physicists' Holy Grail – was that they all predicted the occurrence of a particle with spin 2 and mass 0. In quantum theory *by itself*, that would have been an embarrassment because no such particles are known. But—'

'The graviton, if it exists, has to be a particle with spin 2 and mass 0!' yelled Vikki, caught up in the excitement.

'Precisely. String theory seemed to be telling physicists that if they wanted an effective theory of quantum particles, then gravity would be a necessary consequence. *All* consistent string theories include gravitons.'

'It would have been more satisfactory if only some had,' said Vikki. 'To help pick the right one.'

'True. But it didn't work out that way . . . and it may not be necessary. You see, that People named Edwitten has recently discovered that all five superstring theories fit into a bigger 11D picture, known as *M*-theory.' The Space Hopper noticed Vikki hopping up and down. 'Yes, Vikki, I agree – that's a very imaginative name too, isn't it? But back to my story. There are five different ways to shrink away one of the 11 dimensions of *M*-theory, and they lead to those five 10D superstring theories. So all five competing theories are actually part of the same Big Picture. Maybe it's the Big Picture that really matters.'

He paused. 'There's only one problem with all this stuff', he said. 'Which is a pity, because it's *so* beautiful.'

'What's the problem with it?'

'It may not be true,' said the Space Hopper sadly.

•

'There's no experimental evidence, you see,' he explained. 'Hard to come by – desperately hard to come by. For that matter, the calculations are so difficult that there aren't many theoretical *predictions*, either. So there's precious little to experiment *on*.'

'But couldn't the Planiturthians *tell* if they really lived in a 10D universe?'

'Couldn't the Flatlanders tell that their 2D world was really just part of 3D Spaceland?'

'Touché. Even so, surely there are experiments that would reveal a whole six extra dimensions? For that matter, why didn't they notice them ages ago?'

'Because, Vikki, they didn't have a HyperZoom. Think about a sheet of paper. What's its dimensionality?'

'Two.'

'Really?'

'Oh. In Flatland, yes. In Spaceland, say, it would really be 3D, but very thin along the third dimension.'

'Quite. And the Planiturthian universe *looks* like a 4D spacetime, but in Superstring Territory it is really 10D, but very thin along the fifth, sixth, seventh, eighth, ninth, and tenth dimensions.'

'So thin that the Planiturthians didn't notice?'

'So thin, Vikki, that the Planiturthians *can't* notice. The structure is below the Planck length – the smallest size their instruments can detect.'

'That's a pity. Convenient for the theorists, though.'

'Suspiciously so. But there might be other ways to find out whether they're right. Indirect ones. For instance, in *M*-theory there is a class of branes called Dirichlet branes, because they look rather like a surface discovered long before by a People named Peter-dirichlet. They turn out to be black branes – branes from which light can't escape. You can even interpret a superstring with ends as a superstring that forms a closed loop, part of which is covered by a black brane. And by doing that, you can reformulate black holes in terms of intersecting black branes wrapped round inside a 7D space.'

'And what good does all that do?' asked Vikki.

'It makes very accurate predictions about evaporating black holes. But that kind of evidence is just circumstantial. So right now, all anyone can really do is point to the elegance of the theory, and the beautiful way in which it unifies Relativity and quantum theory.'

'Which makes it right?'

'Not at all. Lots of physicists don't like that approach. They say that beauty could be totally irrelevant. And there are lots of other possibilities in the Mathiverse that don't roll up the extra dimensions of spacetime into very tiny shapes, and there are all sorts of other reasons why the Planiturthians wouldn't notice *them*, either . . . Beauty can be a trap, Vikki.

'Anyway, *everything* in the Mathiverse is beautiful, if you get used to it.' The Space Hopper's ∪ faded momentarily, then renewed itself in a dazzling display of teeth. 'None the less, the idea that the extra dimensions are curled up really tightly, so you can observe them only on the extraordinarily small scale of the Planck length, is still the most appealing explanation of why Planiturthians don't observe ten dimensions.' He paused. For once, the bouncy creature seemed at a loss for words. 'Um – you know something?'

'What?'

'Back in Flatland, Vikki, it's the eve of a new century. Your tour of the Mathiverse has reached its climax. You've learned as much as I can teach you and as much as your brain can digest at one sitting . . . I . . . I think it's time I took you home. Don't you?'

Vikki's centroid leaped. *Oh, yes . . . But.* 'I'll miss you, Hopper! But you're right. It's time I returned to my family. Um.'

'Yes?'

Vikki wanted to ask one last question, to make sure she'd understood the essence of the Planiturthian universe. 'You're saying that when the extra dimensions of spacetime get so thin that they're only *one* Planck length across, they become undetectable?'

'That's right.'

'So the only features of the universe that Planiturthians can observe are those that are . . . as thick as *two* short Plancks?'

'You have it exactly,' declared the Space Hopper.

17

FLATTERLAND

The whatever-it-is that moves narrative through the realms of metaspace did whatever-it-does. Victoria Line was going home. Never again, though, would she imagine that just because Flatland *looked* like a plane, it must therefore *be* a plane. In this, her thinking went beyond eccentric Planiturthians who were convinced that because their world looked flat, therefore it *was* flat; it even went beyond the modern Planiturthians who were convinced that because their world looked three-dimensional, therefore it *was* three-dimensional. However, there was still some way to go before she could compete with those Planiturthians who were convinced that they had no idea what kind of space or spacetime they lived in, but were enjoying the challenge of finding out.

'It would be wonderful,' she mused wistfully, 'to return to Flatland and discover it has hidden dimensions that no one had ever suspected. Even if they're rolled up so tiny that nobody can see them. But I don't suppose that will ever happen.'

'I wouldn't be so sure,' said the Space Hopper. 'Look, we're getting near Flatland now.' He waggled his horns thoughtfully. 'I've got an idea. Instead of dropping you straight back into your home territory, why don't we heave to somewhere nearby and see what it looks like from there?'

'Nearby?'

'In the quasi-metric of metaspace. *Conceptually* nearby. Don't worry what that means – come to think of it, I have no idea what it means, I just know how to *do* it. Oh, and switch on the HyperZoom extension.'

'Why?'

'I've just got this feeling that . . . well, we may come across some surprises.'

The glowing plane of Flatland loomed closer. They could tell, because suddenly it was possible to pick out details – first, the great polygonal oceans, then the zigzag edges of huge forests, which quickly resolved into leafy fractal structures with circular cores.

The Space Hopper muttered to himself and made some adjustments to the VUE. The forest slid away and was replaced by a maze of houses and streets.

They zoomed ever closer.

'The Palace of the Prefect!' Vikki yelled, as she saw a complex of shapes that she recognized. 'And – look! *There's my house!*'

The Space Hopper made some further fine adjustments, and the familiar pentagonal house grew until it was all they could see. Vikki hadn't seen it from this angle, of course – except for a split second when she was first whisked away from her own world, and she hadn't been in the mood to appreciate the fine points at that stage – but she'd got so used to the effect of the VUE that she had little trouble recognizing the layout of the rooms.

Grosvenor and the boys were sitting by the fire; Jubilee was bustling about in the kitchen. Vikki felt a deep-seated twinge of guilt. Her mother always threw herself into the housework when something was troubling her – and Vikki knew very well what that was. Their daughter had gone off mysteriously, without leaving a trace, and (despite the very best of intentions, Vikki told herself) had been too preoccupied with herself to get in touch . . .

She was so busy feeling guilty that it took her quite a long time to notice that there was something very strange about her mother. There was a clear, solid line segment, bustling about in the plane of Flatland . . . but *above* it, pointing along another direction entirely, was a ghostly pentagonal shape. Her mother's lineal features were one of its edges: the rest pointed *out of Flatland altogether.*

Vikki stared at the ghostly apparition, and then at the Space Hopper. 'You *knew* about this,' she said. 'What, exactly, is going on here?'

'Shadow matter,' said the Space Hopper. 'Flatlanders think that their males are two-dimensional polygons, but their females are "only" one-dimensional lines. Quite why they place such importance on apparent dimensionality, beats me – but social hierarchies are usually absurd. However, what neither the males nor the females realize is that Flatland is supersymmetric. It's one of the most basic social symmetries – male/female duality. As different as even and odd, bosons and fermions. But there is a social *super* symmetry, unnoticed until now. Flatland females extend into a third dimension – not *the* third dimension in a Spaceland sense, just a dimension that's different from the ordinary two spatial dimensions of Flatland. Along that dimension, matter has different properties – reflected in the difference in mental outlook between Flatland males and Flatland females. The shadow world and the ordinary world of Flatland meet along lines – not just one line, but *all* lines . . . one reason why the Spaceland model is inadequate. Female consciousness has evolved to recognize the ordinary world, but not the shadow world. But the female *subconscious* has always known about the shadow world.'

Vikki stared at him. 'And you say *all* Flatland females have this extension into the shadow world?'

'Yes.'

'Even – even *me*?' Her voice had gone all squeaky at the implications.

'Of course. You're a Flatland female, right?'

'What – what shape am I really?'

The Space Hopper fiddled with the VUE. 'Mmm, must be a setting for a shadow-matter mirror somewhere . . . not that it's something I often *use*, you see . . . not much call for it most of the time. Where did I put the manual? Oh, yes, tucked away inside the tenth dimension, here it is! Right, see for yourself.'

A ghostly image appeared in front of her VUE-enhanced senses. It was—

'An *octagon*?'

'Very much so. I don't know what your father would think, from that hidebound position of his in which the number of sides determines social position. Your mother is a pentagon – one step higher than him. And you are *much* higher in the social hierarchy.'

Vikki thought about this. 'Not really. The males determine what counts. They'd just say that shadow dimensions don't count.'

The Space Hopper bobbed in agreement. 'Indeed they would. So?'

'So? So what?'

'Exactly.'

'Sorry?'

'So: *what are you going to do about it?*'

Vikki's mind was whirling. Obviously it wouldn't do any good to explain any of this to the men. They'd just declare her insane, and she'd get the modern equivalent of what had happened to poor old Albert.

But – she could tell the *women.*

The only problem was, *how*?

And it was then that the idea came to her.

•

Threeday 1 Angulary 2100

Well, Diary Dear – that was quite a homecoming! I suppose it was a bit of a shock when I suddenly materialized in the living room! Bang on the stroke of midnight at the century's end! Talk about high drama . . .

I don't think I can write down my feelings. Or anybody else's. Some things just don't translate into print.

I chose the dramatic entrance because – well, because I'm *like* that, but mostly because it was the best way to make my story sound credible.

Because Grosvenor and the boys were there, it wasn't possible for Mum and me to have a woman-to-woman talk about important things like the Shadow World . . .

Silly name. Makes it sound unreal. So I've come up with a better one.

Flatterland.

Now, I know what you'll say, Diary my ever-present critic, because you're a pedantic little so-and-so: since Flatland is a plane, which is about as flat as you can get, how can anything be *flatter*? But that's not my meaning. As the Space Hopper said, technically speaking, Flatterland *is* flat – something to do with zero geodesic curvature and other technobabble. But even so, that doesn't make it *flatter*. Obviously.

But flatness isn't the only place where this is a problem. Take 'wetter'. Look: either you're wet or you're not. The *extent* to which you are wet may be lesser or greater, but not your actual state of wetness or dryness.

My dictionary says that you add '-er' to a word to indicate the *comparative*: one interpretation is that X-er is 'like X, only more so'. And that's what Flatterland is. Like Flatland, only more so. The 'more so' is what the Space Hopper called the Shadow World.

So, my idea has a name. In effect, I have a Brand. And in today's market-oriented world, branding opens many doors.

Not, however, the door of distribution. Or dissemination, since what I have to sell is a concept, a Philosophy, a Big Idea. An idea that, given time to take root, will empower all Flatland women. Like the Space Girls said I should. And replace Flatland by Flatterland in *everyone*'s minds – even our shadow-blind males.

Distribution, that's my problem . . .

I can't just wander out on to the streets preaching the Gospel of Flatterland. It would be poor old Albert all over again, in spades. But – this is the age of electronic communications. Which the young understand much better than their elders, and use much more. And the youth of Flatland will be my shock-troops . . .

Ah, now you see it too, Diary.

All it needs is a smattering of HyperDot Mark-up Language and an Anonymous InterLine Link. Then, I'm in business!

•

Jubilee stared at the screen of Vikki's computer. She couldn't understand why her daughter had been so insistent that she should learn how to use the InterLine.

Vikki had gone out, leaving her mother with a scroll of detailed instructions about which buttons to press and so forth. Jubilee had started tentatively, but now she was feeling distinctly proud of herself, for she had successfully gone on-Line and was even now surfing a variety of sites that her daughter had suggested.

An advertisement caught her eye.

WOMEN'S TALK CHAT-POINT.

She looked again at the scroll. Yes, Vikki had listed this as a really interesting site. Suggested she tell Grosvenor all about it. Well, she'd tried to raise the matter, but her husband didn't really seem interested in such things.

She positioned the cursor and clicked.

What she had learned to call an 'interval' opened up. Within were several subintervals – COSMETICS was one, and DRESSES another. There were headings for SHOPPING and SLIMMING and HOW TO MAKE YOUR POLYGON NOTICE YOU and HOW TO DUMP YOUR POLYGON AND BE YOUR OWN LINE, and lots more. About two-thirds of the way down the list was an item on Vikki's scroll: GLOBAL GOSSIP.

Click.

Whatever Jubilee had been expecting, it wasn't what she got. The next screen flashed a brief message: IS THE POLYGON IN YOUR LIFE IN THE HOUSE?

She typed 'NO'.

The screen went blank for a second, while the computer downloaded something called a SQUIF file, and then—

FLATTERLAND, the heading said, in big, antique lettering. Then, in smaller type: *Supersymmetric Sister! The shadow world of gender equality awaits. Superiority, even. If you want to raise your station in Life, click here and be empowered.*

For some reason, the thought of being empowered appealed to Jubilee – even though she wasn't quite sure what it meant. She hesitated, then—

Click.

And a new universe opened.

THE TENTH DIMENSION

Seen from space . . .

But it *was* a space. Well, a spacetime. Start again.

Seen from a ten-dimensional supermanifold, it was a strange world, with the austere beauty of a page from Einstein. In fact, it *was* a page from Einstein, geometry made flesh. A sprawling, humming world of three-dimensional shapes stacked together along one-dimensional time: women, men, infants, toddlers, adolescents . . . People, of their own kind. They lived Peoplish lives, ate Peoplish food, drank Peoplish drinks, made Peoplish love, bore Peoplish children, and died (Peoplishly) in a 3+1-dimensional universe, and never thought it the least bit curious. Their relativistic spacetime continuum was all they could see, all they could hear, all they could feel. To them, it was all there was.

As long as nothing disturbed that perception, it was *true*.

But times (and spaces) were changing in Spacetimeland . . .

INDEX